RICHARD VON

Probability, Statistics and Truth

SECOND REVISED ENGLISH EDITION
PREPARED BY
HILDA GEIRINGER

DOVER PUBLICATIONS, INC.
NEW YORK

Published in Canada by General Publishing Company, Ltd.,
30 Lesmill Road, Don Mills, Toronto, Ontario.
Published in the United Kingdom by Constable and Company,
Ltd.

This Dover edition, first published in 1981, is an unabridged
republication of the second revised English edition originally
published in 1957 by George Allen & Unwin Ltd., London. The
present edition is published by special arrangement with George
Allen & Unwin Ltd.
The work was originally published in German by J. Springer in
1928; the third German edition, 1951, is the definitive version
and the basis of the present English edition (a revision of the
English translation of 1939).

International Standard Book Number: 0-486-24214-5
Library of Congress Catalog Card Number: 81-67040

Manufactured in the United States of America
Dover Publications, Inc.
180 Varick Street
New York, N.Y. 10014

PREFACE

THE present second English edition is not merely a reprinting of the first edition, which has been out of print for several years, but rather a translation of the third German edition, revised by the author in 1951; this last differs in many ways from the second German edition, on which the original English translation was based.

The change consists essentially in the author's omitting some of the discussions of the early controversies regarding his theory, and making instead various additions: The concept of randomness, which plays a central role in the author's theory, is reconsidered—in particular, with respect to the problem of mathematical consistency—and carefully reformulated. The question of substituting for it some 'limited randomness' is taken up, and the author concludes that, as far as the basic axioms are concerned, no such restriction is advisable. Systematic consideration is given to recent work concerned with the basic definitions of probability theory: in the ideas of E. Tornier and of J. L. Doob, the author sees a remarkable development of his theory of frequencies in collectives. The analysis of the two Laws of Large Numbers has often been considered an outstanding section of the book; in the 1951 edition, on the basis of new mathematical results, the discussion of the Second Law is deepened and enlarged, with the aim of clarifying this highly controversial subject. Comments are added on the testing of hypotheses, (as based on the inference theory originated by T. Bayes), on R. A. Fisher's 'likelihood', and on a few related subjects. These and all other additions are selected as well as discussed in relation to the basic ideas advanced in the book.

The deviations in content of the present English version from the third German edition are insignificant. Some passages which seemed of more local (Austrian) interest have been omitted. In a few instances the text has been changed (with explanation in the Notes when necessary). The Notes, historical, bibliographical, etc., have been somewhat modified. References have been brought up to date; indications of German translations have been replaced by corresponding English works; several notes are new. A subject index has been added.

The present text is based on the excellent translation (1939) of

Messrs. J. Neyman, D. Scholl, and E. Rabinowitsch; it has been supplemented by all the new material in the 1951 edition, and amended in the light of notes made by the author in anticipation of a new English edition. In addition, the entire text was given a careful editorial revision. The sixth chapter, on 'Statistics in Physics' was essentially retranslated.

In these various aspects, and in particular with regard to the new translations, I enjoyed the valuable assistance of Mrs. R. Buka. My sincere thanks go to Professor J. Neyman for his understanding encouragement, to Professor E. Tornier for significant advice regarding a few difficult passages, to Dr. A. O'Neill who prepared the index, to Mr. F. J. Zucker, who kindly read the translation of the sixth chapter, and to Mr. J. D. Elder, who was good enough to check the text with respect to uniformity and general consistency of style. I am very grateful to the Department of Mathematics, particularly to Professor G. Birkhoff, and to the Division of Engineering and Applied Physics of Harvard University, who together with the Office of Naval Research sponsored this work. Finally, I thank the publishers, Allen & Unwin, Ltd., London, and the Macmillan Company, New York, for the cooperation they gave freely whenever needed, and the Springer-Verlag, Vienna, for granting permission to bring out this edition.

Cambridge, Mass., U.S.A. HILDA GEIRINGER
June 1956.

PREFACE TO THE THIRD GERMAN EDITION

THIRTY years have gone by since the first publication of this book. The present sixth edition follows two in German, and one each in Russian, English and Spanish. I therefore feel that it may be well to review briefly at this point the development of the ideas with which we deal here.

The subject of this book is the quantitative concept of probability in probability theory, a discipline which originated in the seventeenth century and which was first set down in a precise manner by Laplace, shortly after 1800. As in other branches of science, such as geometry, mechanics, and parts of theoretical physics, so in the theory of probability, the epistemological position remained in the dark for a long time. Near the end of the nineteenth century, Ernst Mach and Henri Poincaré made decisive contributions towards the clarification of the meaning and purpose of scientific concepts. However, Mach was not interested in probability, and Poincaré accepted quite uncritically the point of view of Laplace, who had a laissez-faire attitude: Beginning with a few not very meaningful words concerning equally possible, favourable and unfavourable cases, he deduced some simple rules which he then used (apparently) to an extent quite out of proportion to his modest starting point. Neither Laplace nor any of his followers, including Poincaré, ever reveals how, starting with *a priori* premises concerning equally possible cases, the sudden transition to the description of real statistical events is to be made.

The essentially new idea which appeared about 1919 (though it was to a certain extent anticipated by A. A. Cournot in France, John Venn in England, and Georg Helm in Germany) was to consider the theory of probability as a science of the same order as geometry or theoretical mechanics. In other words, to maintain that just as the subject matter of geometry is the study of space phenomena, so probability theory deals with mass phenomena and repetitive events. By means of the methods of abstraction and idealization (which are only in part free activities of the mind), a system of basic concepts is created upon which a logical structure can then be erected. Owing to the original relation between the basic concepts and the observed primary phenomena, this theoretical structure permits us to draw conclusions concerning the world of reality. In order to allow a rationally justified application of this probability theory to reality, a quantitative probability concept must be defined in terms of potentially unlimited sequences of observations or experiments. The relative frequency of the repetition is the 'measure' of probability, just as the length of a column of mercury is the 'measure' of temperature.

These are the fundamental ideas underlying the new concept; when they were first formulated they appeared to break completely with the views generally held; today, however, they have been accepted, in the main, by most authors in the field.

This development was somewhat masked by the fact that these same thirty years, starting about 1918, witnessed progress on an unprecedented scale on the formal side of the *mathematics* of probability. My first modest attempt to arrive at certain general formulations ('Fundamentalsätze der Wahrscheinlichkeitsrechnung', *Mathematische Zeitschrift*, 1918) is today, in most respects, outdated. With the aid of the modern theory of sets and the theory of real functions, it has been possible to perfect the formal mathematical foundation in a manner that is aesthetically satisfying and not without practical value. This is true especially for some fairly recent results which have opened up the important field of so-called stochastic processes. However, a discussion of these topics does not belong in this book. Suffice it to say that the detailed development of the tautological side of a discipline affects neither the need for, nor the content of its epistemological foundation. The mathematician who works on special problems of higher geometry need not concern himself with the axiomatic aspect of Euclid's Elements; this, however, does not imply that there is any part of geometry that is not based on the Euclidian elements.

A certain reaction—of a rather superficial nature—has set in during the last few years. It is brought out repeatedly in this book that the word 'probability' has a meaning in everyday language that is different from its quantitative meaning in probability calculus. Some authors with metaphysical leanings have sought to build a separate theory on this other meaning of the word. Such attempts, namely, the study of questions of reliability or plausibility of judgements, of propositions and systems of propositions, are justified as long as they remain within certain limits. However, as soon as numerical values are attributed to these plausibilities and used in calculation, one has either to accept the frequency definition of probability (as is done by some authors) or to fall back on an a priori standpoint based on equally likely cases (as is done by others). The stated purpose of these investigations is to create a theory of

vi

induction or 'inductive logic'. According to the basic viewpoint of this book, the theory of probability in its application to reality is itself an inductive science; its results and formulas cannot serve to found the inductive process as such, much less to provide numerical values for the plausibility of any other branch of inductive science, say the general theory of relativity.

During the last few decades, the Anglo-Saxon countries have seen a good deal of progress in the practical application of probability theory to statistical problems in biology and in industry. This development started with the misunderstanding of one of the classical formulas of probability calculus (Bayes's rule); there followed a period that was characterized by the erroneous practice of drawing statistical conclusions from short sequences of observations: the so-called 'small sample theory'. At present, it seems that these mistakes have essentially been overcome and that the way has been cleared for an appropriate development of those formulas that permit an extensive application of statistical methods. The present edition contains a few remarks on this subject.

Otherwise, this new edition differs from the preceding ones chiefly in the omission of some polemic discussions and criticisms that seemed no longer necessary, while various additions are made, mainly in Lectures 3, 4, and 5.

Cambridge, Mass., U.S.A.   R. v. MISES
November 1950

CONTENTS

CONTENTS

THIRD LECTURE

Critical Discussion of the Foundations of Probability

CONTENTS

CONTENTS

SIXTH LECTURE

Statistical Problems in Physics

The Definition of Probability

To illustrate the apparent contrast between statistics and truth which might be inferred from the title of our book may I quote a remark I once overheard: 'There are three kinds of lies: white lies, which are justifiable; common lies—these have no justification; and statistics.' Our meaning is similar when we say: 'Anything can be proved by figures'; or, modifying a well-known quotation from Goethe, with numbers 'all men may contend their charming systems to defend.'

At the basis of all such remarks lies the conviction that conclusions drawn from statistical considerations are at best uncertain and at worst misleading. I do not deny that a great deal of meaningless and unfounded talk is presented to the public in the name of statistics. But my purpose is to show that, starting from statistical observations and applying to them a clear and precise concept of probability it is possible to arrive at conclusions which are just as reliable and 'truth-full' and quite as practically useful as those obtained in any other exact science. In order to achieve this purpose, I must ask you to follow me along a road which is often laborious and by paths which at first sight may appear unnecessarily winding.

AMENDMENT OF POPULAR TERMINOLOGY

'All our philosophy is a correction of the common usage of words,' says Lichtenberg.[1] Many of the quarrels and mistakes occurring in the course of scientific advance could be avoided if this remark were always remembered. Our first step, therefore, will be to inquire more closely into the meaning of the *word* 'probability'. This will be followed by arguments which will gradually lead us to an adequate scientific definition of the *concept* of probability. I have already hinted that the key to the relation between statistics and truth may be found in a reasonable definition of probability. I hope that this

1

point will become quite clear in the course of the subsequent discussion.

The word 'probable' is frequently used in everyday speech. We say, for instance, 'It will probably rain tomorrow', or, 'It is probably snowing right now in Iceland', or 'The temperature was probably lower a year ago today than it is today'. Again, we speak of something being more or less probable or more or less improbable when, for example, we are discussing the guilt of an accused person or the deposition of a witness. In a more definite way, we may say that there is a greater probability of winning the first prize in a certain sweepstake than of gaining the smallest in another. We have no difficulty in explaining what we mean by these statements as long as the inquirer is satisfied by a 'descriptive' answer. We can easily find a number of expressions which will serve. We may speak of a 'guess', of 'approximate' or 'incomplete' knowledge, or 'chance', or we may say that we have more or less adequate reasons for believing that this or that is the case, and so forth.

EXPLANATION OF WORDS

Considerable difficulties arise, however, when we are asked to give an exact explanation, or, even more, a definition of what we mean by 'probability'. Perhaps someone may suggest looking up the word in a dictionary. Volume XIII of the German Dictionary by Jakob and Wilhelm Grimm[2] gives us detailed information: The Latin term 'probabilis', we are told, was at one time translated by 'like truth', or, by 'with an appearance of truth' ('mit einem Schein der Wahrheit'). Only since the middle of the seventeenth century has it been rendered by 'wahrscheinlich' (lit. truth-resembling). We also find a number of quotations illustrating the use of the word, most of them taken from philosophical works. I shall only refer to a few examples: 'The probable is something which lies midway between truth and error' (Thomasius, 1688); 'An assertion, of which the contrary is not completely self-contradictory or impossible, is called probable' (Reimarus). Kant says: 'That which, if it were held as truth, would be more than half certain, is called probable.' Perhaps, after these examples, someone may wish to know what modern philosophy has contributed to this subject. I quote literally from Robert Eisler's *Dictionary of Philosophic Concepts* (1910): 'Probability, in the subjective sense, is a degree of certainty which is based on strong or even overwhelming reasons for making an assertion. . . . In the objective sense, the probable is that which is supported by a number of objective arguments. . . . There are differing degrees of probability and these depend upon the kind and number of reasons or facts upon which the assertion or conclusion of probability is based.'[3]

To consider now a familiar, modern source, Webster's *New International Dictionary* gives the following definition of probability: 'Quality or state of being probable; reasonable ground for presuming; likelihood; more narrowly, a conclusion that is not proof but follows logically from such evidence as is available; as, reports devoid of all *probability*, to establish *probability* of guilt.'[4] (For Webster's definition of mathematical probability, see note 1, Lect. 3).

It is useless to quarrel with these philosophic explanations. They are merely substitutions; one word is replaced by others and frequently by a great many. If these new words are more familiar to the reader than the original one, then he may find some explanation in this procedure while others will find none in this way. Some, for instance, may understand the meaning of 'more than half certain' better than the simple word 'probable'. This can only be a matter of personal preference, and explanations of this kind cannot be generally regarded as a correction of common word usage.

SYNTHETIC DEFINITIONS

Let us now consider a way by which we may arrive at a better definition of probability than that given in the dictionaries, which is so obviously unsatisfactory for our purpose.

In the course of the last few centuries a method of forming and defining concepts has been developed by the exact sciences which shows us the way clearly and with certainty. To ignore or to reject this method would be to question all the achievements of modern mathematics and physics. As a preliminary example, let me quote a modern definition of a concept which belongs to the science of sociology; this is more nearly related to the subject-matter of our general education and will thus form a transition to those concepts with which we shall be concerned later. Werner Sombart,[5] in his book *Proletarian Socialism*, attempts to find a useful definition of his subject and in so doing he considers a number of current interpretations. He concludes: 'The only remaining possibility is to consider socialism as an idea and to form a reasonable concept of it, i.e., to delimit a subject matter which possesses a number of characteristics considered to be particularly important to it and which form a meaningful unity; the "correctness" of this concept can only be judged from its fruitfulness, both as a creative idea in life and as a useful instrument for advancing scientific investigation.' These words actually contain almost all that is characteristic of the scientific method of developing new concepts. There are in particular two points which I wish to emphasize: in the first place, the content of a concept is not derived from the meaning popularly given to a word, and it is therefore independent of current usage. Instead, the concept is first established and its boundaries are purposely circumscribed, and a word, as a suitable kind of label, is affixed later. In the second place, the value of a concept is not gauged by its correspondence with some usual group of notions, but only by its usefulness for further scientific development, and so, indirectly, for everyday affairs.

We may say, with Kant,[6] that our aim is to give not an analytic definition of probability but a synthetic one. We may leave open the question of the general possibility of finding analytic definitions at all.

TERMINOLOGY

I should like to add a further remark about the first of the above-mentioned properties of synthetic definitions. The person who arrives at a new scientific concept may be inclined to invent a *new* name for it: he will look for a word which has not already been used in some other sense, perhaps one closely related to that in which he himself wishes to use it. Since it is obviously difficult to find new words in one's own language, foreign or adopted words are frequently introduced into the scientific vocabulary. Although the purists in the matter of language are not altogether to be blamed, it would appear that they go too far when they ignore this reason for the introduction of foreign words into the language of science, and attempt to retranslate them into ordinary language. For example, it is unfortunate that most languages have no specific word for probability in its scientific sense but only popular terms like Wahrscheinlichkeit, probability, probabilité. However, no term has been invented and, naturally, it is quite possible for a scientific concept to exist without having a special name. This is the case with many of the most important concepts of mechanics which are hidden behind such ordinary words as force, mass, work, etc. All the same, I do feel that many laymen, and even some professionals in the field of mechanics, would understand these concepts more clearly if they had Latin names rather than names taken from everyday usage. Scientists themselves are only human, and they use the common language for the greater part of their lives in the same way as other humans. They are subject to the same kinds of confusion of speech and often give way to them only too freely.

THE CONCEPT OF WORK IN MECHANICS

Before I deal with the development of the scientific concept of probability, I should like to recall the similar state of affairs which prevailed during the formation of most other concepts in the exact sciences. As an example of a concept which is familiar today to educated persons, I shall choose that of *work* as it is used in theoretical mechanics. We all use the word 'work' with a variety of different meanings. Even when we do not consider idiomatic phrases like 'to work on someone's feelings', there are many ways in which the word is used which have little to do with the concept of work as it is understood in science. The simplest scientific definitions of work are: 'Work is the product of force and distance', or, more exactly, 'the scalar product of the vectors of force and displacement', or 'the line-integral of force'.[7] All these definitions are equally suitable for many everyday matters and the nonmathematician need only keep in mind the first of them. If we consider some examples, such as the lifting of a weight, the turning of a crank, or the pushing of a pedal, in each case the work performed becomes greater with an increase in the weight of the load moved as well as with an increase in the distance through which it is moved.

Yet this scientific definition of work is hardly applicable to even the simplest of activities which are only partly of a mechanical nature. We may think of working a typewriter or playing a musical instrument. In the latter case, it is hardly possible to say that the correct measure of the work performed by the musician is the product of the force applied by the fingers of the musician and their displacement. Again, when we speak of the work involved in writing a book, painting a picture, or attending a patient, we are even further from the scientific meaning of the word 'work'. It is hard work from the human point of view to hold a heavy weight steadily with outstretched arms, but in this case the product of the force and the displacement is zero. In sports and games, the work calculated according to the rules of mechanics can hardly be regarded as a correct measure of the physical effort involved. No reasonable person objects to these discrepancies because we have become too accustomed to the fact that the same word may have a different

5

meaning according as it is used scientifically or colloquially. When we use the word 'work' in its scientific meaning, we automatically eliminate all other associations which it may bring to our minds on other occasions, since these do not appertain to it in mechanics.

AN HISTORICAL INTERLUDE

It was not immediately realized that the meaning of scientific concepts is independent of the literal meanings of the words used for them; this recognition only evolved over a long period in the development of scientific thought. This is illustrated by the controversy between the followers of Descartes and of Leibnitz on the question of *vis viva*.[8] Is the 'effect of a force' equal to the product of the mass and the velocity or to the product of the mass and half the square of the velocity? We know now that this question has no logical answer and relies upon a definition which is ultimately arbitrary; what is to be called '*vis viva*' and what '*momentum*' is completely secondary. In the words of Robert Meyer we may say: 'It does not matter what others mean by the word "work", what we intend to convey by it is the thing that really matters'.

We have all experienced, in school, the difficulties which arise from the confusion between the colloquial and the scientific meanings of words. We had to *learn* that even the slowest motion has velocity, that a retarded motion has an acceleration but with a negative sign, and that 'rest' is a particular case of motion. This mastering of scientific language is essential in mental development, for, without it, there is no approach to modern natural science.

We have given a few examples of the use of common words as scientific terms. There is a growing tendency towards a more precise use of certain words of everyday language. Most educated persons are nowadays aware of the specific meanings given to words in science. They may, for instance, be expected to distinguish between the words quadrangle, rectangle, and square, and to know how these three terms are defined. In the near future, a much deeper understanding of questions of this kind will be taken for granted.

THE PURPOSE OF RATIONAL CONCEPTS

When a name is chosen for a scientific concept, it is obvious that we must consider linguistic convenience and good taste. Nevertheless, it is the content of a concept, and not its name, which is of importance. The definition must serve a useful purpose. We consider a purpose to be useful if it is in agreement with what we generally regard as the purpose of science. This is to bring order into the multiplicity of observed phenomena, to predict the course of their development, and to point out ways by which we may bring about particular phenomena in which we are interested. The scientific notion of 'work', and the whole conceptual system of classical physics, of which this notion is a part, have proved their utility in all these directions. The Law of Conservation of Energy has provided us with the means of bringing order into a very wide region of physical phenomena. This law enables us to predict the course of many natural events, while, at the same time, the engineer and the electrician derive from it the data necessary to calculate the dimensions of their machines. Nobody can deny the theoretical and practical success of scientific mechanics, which has been founded upon concepts of this kind. One criticism occasionally levelled against the practical utility of this rationalization of scientific concepts will be examined briefly.

People have said, 'It is easy to formulate a self-consistent theory based on exactly defined artificial concepts, but in the practical applications of the theory we always have to deal with vague processes which can only be adequately described in terms of correspondingly vague concepts which have evolved in a natural way'. There is some truth in this objection, for it makes evident a great deficiency which is to be found in any theoretical treatment of reality. The events which we observe, and in which we take part, are always very complicated; even the most elaborate and detailed theory cannot take into account all the factors involved. It is an art requiring a scientifically trained mind merely to identify the one feature among a multitude present in a natural process which is to be considered as the only essential one from the theoretical point of view. Nevertheless, it would be a mistake, or at least would lead us away from the whole of the scientific development of the last few centuries, if we were to follow the Bergson school of modern philosophy. The adherents of this school repudiate the use of sharply defined concepts, hoping in this way to cope more adequately with the complexity of the real world. Nothing would be gained by a return to those vague notions, which are sometimes praised as intuitive but which are really nothing but an unprecise and indefinite use of words.

THE INADEQUACY OF THEORIES

Imagine that I draw a 'straight line' on a blackboard with a piece of chalk. What a complicated thing is this 'line' compared with the

7

'straight line' defined by geometry! In the first place, it is not a line at all, since it has definite breadth; even more than that, it is a three-dimensional body made of chalk, an aggregate of many small bodies, the chalk particles. A person who was unaccustomed to seeing the teacher at school draw 'straight lines' of this kind would be almost unable to understand what this body of chalk has in common with the 'straight line' defined in the textbooks as 'the shortest distance between two points'. All the same, we do know that the exact idealized conceptions of pure geometry are essential tools for dealing with the real things around us. We need these abstract concepts just because they are simple enough that our minds can handle them with comparative ease.

Attempts have been made to construct geometries in which no 'infinitely narrow' lines exist but only those of definite width. The results were meagre because this method of treatment is much more difficult than the usual one. Moreover, a strip of definite width is only another abstraction no better than a straight line, and is really more complicated, since it involves something like two straight lines limiting it, one on either side.

I am prepared to concede without further argument that all the theoretical constructions, including geometry, which are used in the various branches of physics are only imperfect instruments to enable the world of empirical fact to be reconstructed in our minds. The theory of probability, which we include among the exact sciences, is just one such theoretical system. But I do not believe that there is any other way to achieve progress in science than the old method: to begin with the simplest, i.e., the exact theoretical scheme and to extend and improve it gradually. In dealing with the theory of probability, i.e., with probability calculus, I do not hope to achieve more than the results already attained by geometry, mechanics, and certain other branches of physics. That is to say, I aim at the construction of a rational theory, based on the simplest possible exact concepts, one which, although admittedly inadequate to represent the complexity of the real processes, is able to reproduce satisfactorily some of their essential properties.

LIMITATION OF SCOPE

After all these preliminary discussions, we now come to the description of our concept of probability. It follows from our previous remarks that our first task must be one of elimination. From the complex of ideas which are colloquially covered by the word 'probability', we must remove all those that remain outside the theory we are endeavouring to formulate. I shall therefore begin with a preliminary delimitation of our concept of probability; this will be developed into a more precise definition during the course of our discussion.

Our probability theory has nothing to do with questions such as: 'Is there a probability of Germany being at some time in the future involved in a war with Liberia?' Again, the question of the 'probability' of the correct interpretation of a certain passage from the *Annals* of Tacitus has nothing in common with our theory. It need hardly be pointed out that we are likewise unconcerned with the 'intrinsic probability' of a work of art. The relation of our theory to Goethe's superb dialogue on *Truth and Probability in Fine Art*[9] is thus only one of similarity in the sounds of words and consequently is irrelevant. We shall not deal with the problem of the historical accuracy of Biblical narratives, although it is interesting to note that a Russian mathematician, A. Markoff,[10] inspired by the ideas of the eighteenth-century Enlightenment, wished to see the theory of probability applied to this subject. Similarly, we shall not concern ourselves with any of those problems of the moral sciences which were so ingeniously treated by Laplace[11] in his *Essai Philosophique*. The unlimited extension of the validity of the exact sciences was a characteristic feature of the exaggerated rationalism of the eighteenth century. We do not intend to commit the same mistake.

Problems such as the probable reliability of witnesses and the correctness of judicial verdicts lie more or less on the boundary of the region which we are going to include in our treatment. These problems have been the subject of many scientific discussions; Poisson[12] chose them as the title of his famous book.

To reach the essence of the problems of probability which do form the subject-matter of this book, we must consider, for example, the probability of winning in a carefully defined game of chance. Is it sensible to bet that a 'double 6' will appear at least once if two dice are thrown twenty-four times? Is this result 'probable'? More exactly, how great is its probability? Such are the questions we feel able to answer. Many problems of considerable importance in everyday life belong to the same class and can be treated in the same way; examples of these are many problems connected with insurance, such as those concerning the probability of illness or death occurring under carefully specified conditions, the premium which must be asked for insurance against a particular kind of risk, and so forth.

Besides the games of chance and certain problems relating to social mass phenomena, there is a third field in which our concept has a useful application. This is in the treatment of certain mechanical and physical phenomena. Typical examples may be seen in the movement of molecules in a gas or in the random motion of colloidal particles which can be observed with the ultramicroscope. ('Colloid' is the name given to a system of very fine particles freely suspended in a medium, with the size of the particles so minute that the whole appears to the naked eye to be a homogeneous liquid.)

UNLIMITED REPETITION

What is the common feature in the last three examples and what is the essential distinction between the meaning of 'probability' in these cases and its meaning in the earlier examples which we have excluded from our treatment? One common feature can be recognized easily, and we think it crucial. In games of chance, in the problems of insurance, and in the molecular processes we find events repeating themselves again and again. They are mass phenomena or repetitive events. The throwing of a pair of dice is an event which can theoretically be repeated an unlimited number of times, for we do not take into account the wear of the box or the possibility that the dice may break. If we are dealing with a typical problem of insurance, we can imagine a great army of individuals insuring themselves against the same risk, and the repeated occurrence of events of a similar kind (e.g., deaths) are registered in the records of insurance companies. In the third case, that of the molecules or colloidal particles, the immense number of particles partaking in each process is a fundamental feature of the whole conception.

On the other hand, this unlimited repetition, this 'mass character', is typically absent in the case of all the examples previously excluded. The implication of Germany in a war with the Republic of Liberia is not a situation which frequently repeats itself; the uncertainties that occur in the transcription of ancient authors are, in general, of a too individual character for them to be treated as mass phenomena. The question of the trustworthiness of the historical narratives of the Bible is clearly unique and cannot be considered as a link in a chain of analogous problems. We classified the reliability and trustworthiness of witnesses and judges as a borderline case since we may feel reasonable doubt whether similar situations occur sufficiently frequently and uniformly for them to be considered as repetitive phenomena.

10

We state here explicitly: The rational concept of probability, which is the only basis of probability calculus, applies only to problems in which either the same event repeats itself again and again, or a great number of uniform elements are involved at the same time. Using the language of physics, we may say that in order to apply the theory of probability we must have a practically unlimited sequence of uniform observations.

THE COLLECTIVE

A good example of a mass phenomenon suitable for the application of the theory of probability is the inheritance of certain characteristics, e.g., the colour of flowers resulting from the cultivation of large numbers of plants of a given species from a given seed. Here we can easily recognize what is meant by the words 'a repetitive event'. There is primarily a single instance: the growing of one plant and the observation of the colour of its flowers. Then comes the comprehensive treatment of a great number of such instances, considered as parts of one greater unity. The individual elements belonging to this unity differ from each other only with respect to a single attribute, the colour of the flowers.

In games of dice, the individual event is a single throw of the dice from the box and the attribute is the observation of the number of points shown by the dice. In the game 'heads or tails', each toss of the coin is an individual event, and the side of the coin which is uppermost is the attribute. In life insurance the single event is the life of the individual and the attribute observed is either the age at which the individual dies or, more generally, the moment at which the insurance company becomes liable for payment. When we speak of 'the probability of death', the exact meaning of this expression can be defined in the following way only. We must not think of an individual, but of a certain class as a whole, e.g., 'all insured men forty-one years old living in a given country and not engaged in certain dangerous occupations'. A probability of death is attached to this class of men or to another class that can be defined in a similar way. We can say nothing about the probability of death of an individual even if we know his condition of life and health in detail. The phrase 'probability of death', when it refers to a single person, has no meaning at all for us. This is one of the most important consequences of our definition of probability and we shall discuss this point in greater detail later on.

We must now introduce a new term, which will be very useful

11

during the future course of our argument. This term is 'the collective', and it denotes a sequence of uniform events or processes which differ by certain observable attributes, say colours, numbers, or anything else. In a preliminary way we state: All the peas grown by a botanist concerned with the problem of heredity may be considered as a collective, the attributes in which we are interested being the different colours of the flowers. All the throws of dice made in the course of a game form a collective wherein the attribute of the single event is the number of points thrown. Again, all the molecules in a given volume of gas may be considered as a collective, and the attribute of a single molecule might be its velocity. A further example of a collective is the whole class of insured men and women whose ages at death have been registered by an insurance office. The principle which underlies the whole of our treatment of the probability problem is that a collective must exist before we begin to speak of probability. The definition of probability which we shall give is only concerned with 'the probability of encountering a certain attribute in a given collective'.

THE FIRST STEP TOWARDS A DEFINITION

After our previous discussion it should not be difficult to arrive at a rough form of definition of probability. We may consider a game with two dice. The attribute of a single throw is the sum of the points showing on the upper sides of the two dice. What shall we call the probability of the attribute '12', i.e., the case of each die showing six points? When we have thrown the dice a large number of times, say 200, and noted the results, we find that 12 has appeared a certain number of times, perhaps five times. The ratio $5/200 = 1/40$ is called the frequency, or more accurately the relative frequency, of the attribute '12' in the first 200 throws. If we continue the game for another 200 throws, we can find the corresponding relative frequency for 400 throws, and so on. The ratios which are obtained in this way will differ a little from the first one, $1/40$. If the ratios were to continue to show considerable variation after the game had been repeated 2000, 4000, or a still larger number of times, then the question whether there is a definite probability of the result '12' would not arise at all. It is essential for the theory of probability that experience has shown that in the game of dice, as in all the other mass phenomena which we have mentioned, the relative frequencies of certain attributes become more and more stable as the number of observations is increased. We shall discuss the idea of 'the limiting value of the relative frequency' later on; meanwhile, we assume that the frequency is being computed with a limited accuracy only, so that small deviations are not perceptible. This approximate value of the relative frequency we shall, preliminarily, regard as the probability of the attribute in question, e.g., the probability of the result '12' in the game of dice. It is obvious that if we define probability in this way, it will be a number less than 1, that is, a proper fraction.

TWO DIFFERENT PAIRS OF DICE

I have here two pairs of dice which are apparently alike. By repeatedly throwing one pair, it is found that the relative frequency of the 'double 6' approaches a value of 0.028, or 1/36, as the number of trials is increased. The second pair shows a relative frequency for the '12' which is four times as large. The first pair is usually called a pair of true dice, the second is called biased, but our definition of probability applies equally to both pairs. Whether or not a die is biased is as irrelevant for our theory as is the moral integrity of a patient when a physician is diagnosing his illness. 1800 throws were made with each pair of these dice. The sum '12' appeared 48 times with the first pair and 178 times with the second. The relative frequencies are

$$\tfrac{48}{1800} = \tfrac{1}{37.5} = 0.027$$

and

$$\tfrac{178}{1800} = \tfrac{1}{10.1} = 0.099.$$

These ratios became practically constant towards the end of the series of trials. For instance, after the 1500th throw they were 0.023 and 0.094 respectively. The differences between the values calculated at this stage and later on did not exceed 10–15%.

It is impossible for me to show you a lengthy experiment in the throwing of dice during the course of this lecture since it would take too long. It is sufficient to make a few trials with the second pair of dice to see that at least one 6 appears at nearly every throw; this is a result very different from that obtained with the other pair. In fact, it can be shown that if we throw one of the dice belonging to the second pair, the relative frequency with which a single 6 appears is about 1/3, whereas for either of the first pair this frequency is almost exactly 1/6. In order to realize clearly what our meaning of probability implies, it will be useful to think of these two pairs of dice as often as possible; each pair has a characteristic probability of showing 'double 6', but these probabilities differ widely.

Here we have the 'primary phenomenon' (Urphänomen) of the theory of probability in its simplest form. The probability of a 6 is a physical property of a given die and is a property analogous to its mass, specific heat, or electrical resistance. Similarly, for a given *pair of dice* (including of course the total setup) the probability of a 'double 6' is a characteristic property, a physical constant belonging to the experiment as a whole and comparable with all its other physical properties. The theory of probability is only concerned with relations existing between physical quantities of this kind.

LIMITING VALUE OF RELATIVE FREQUENCY

I have used the expression 'limiting value', which belongs to higher analysis, without further explanation.[13] We do not need to know much about the mathematical definition of this expression, since we propose to use it in a manner which can be understood by anyone, however ignorant of higher mathematics. Let us calculate the relative frequency of an attribute in a collective. This is the ratio of the number of cases in which the attribute has been found to the total number of observations. We shall calculate it with a certain limited accuracy, i.e., to a certain number of decimal places without asking what the following figures might be. Suppose, for instance, that we play 'heads or tails' a number of times and calculate the relative frequency of 'heads'. If the number of games is increased and if we always stop at the same decimal place in calculating the relative frequency, then, eventually, the results of such calculations will cease to change. If the relative frequency of heads is calculated accurately to the first decimal place, it would not be difficult to attain constancy in this first approximation. In fact, perhaps after some 500 games, this first approximation will reach the value of 0.5 and will not change afterwards. It will take us much longer to arrive at a constant value for the second approximation, calculated to two decimal places. For this purpose it may be necessary to calculate the relative frequency in intervals of, say, 500 casts, i.e., after the 500th, 1000th, 1500th, and 2000th cast, and so on. Perhaps more than 10,000 casts will be required to show that now the second figure also ceases to change and remains equal to 0, so that the relative frequency remains constantly 0.50. Of course it is impossible to continue an experiment of this kind indefinitely. Two experimenters, co-operating efficiently, may be able to make up to 1000 observations per hour, but not more. Imagine, for example, that the experiment has been continued for ten hours and that the relative frequency remained constant at

0.50 during the last two hours. An astute observer might perhaps have managed to calculate the third figure as well, and might have found that the changes in this figure during the last hours, although still occurring, were limited to a comparatively narrow range.

Considering these results, a scientifically trained mind may easily accept the hypothesis that by continuing this play for a sufficiently long time under conditions which do not change (insofar as this is practically possible), one would arrive at constant values for the third, fourth, and all the following decimal places as well. The expression we used, stating that the relative frequency of the attribute 'heads' tends to a limit, is no more than a short description of the situation assumed in this hypothesis.

Take a sheet of graph paper and draw a curve with the otal number of observations as abscissæ and the value of the relative frequency of the result 'heads' as ordinates. At the beginning this curve shows large oscillations, but gradually they become smaller and smaller, and the curve approaches a straight horizontal line. At last the oscillations become so small that they cannot be represented on the diagram, even if a very large scale is used. It is of no importance for our purpose if the ordinate of the final horizontal line is 0.6, or any other value, instead of 0.5. The important point is the existence of this straight line. The ordinate of this horizontal line is the limiting value of the relative frequency represented by the diagram, in our case the relative frequency of the event 'heads'.

Let us now add further precision to our previous definition of the collective. We will say that a collective is a mass phenomenon or a repetitive event, or, simply, a long sequence of observations for which there are sufficient reasons to believe that the relative frequency of the observed attribute would tend to a fixed limit if the observations were indefinitely continued. This limit will be called *the probability of the attribute considered within the given collective*. This expression being a little cumbersome, it is obviously not necessary to repeat it always. Occasionally, we may speak simply of the *probability of 'heads'*. The important thing to remember is that this is only an abbreviation, and that we should know exactly the kind of collective to which we are referring. 'The probability of winning a battle', for instance, has no place in our theory of probability, because we cannot think of a collective to which it belongs. The theory of probability cannot be applied to this problem any more than the physical concept of work can be applied to the calculation of the 'work' done by an actor in reciting his part in a play.

THE EXPERIMENTAL BASIS OF THE THEORY OF GAMES

It will be useful to consider how the fundamental experiment of the determination of probability can be carried out in the other two cases mentioned: I mean in the case of life insurance, and that of molecules of a gas. Before doing this, I should like to add a few more words on the question of games of chance. People may ask, 'How do we know for certain that a game of chance will develop in practice along the lines which we have discussed, i.e., tending towards a stabilization of the relative frequencies of the possible results? Is there a sufficient basis for this important assumption in actual sequences of experiments? Are not all experiments limited to a relatively short initial stage?' The experimental material is, however, not as restricted as it may appear at first sight. The great gambling banks in Monte Carlo and elsewhere have collected data relating to many millions of repetitions of one and the same game. These banks do quite well on the assumption of the existence of a limiting value of the relative frequency of each possible result. The occasional occurrence of 'breaking the bank' is not an argument against the validity of this theory. This could only be questioned on the basis of a substantial decrease in the total earnings of the bank from the beginning of its operation to any specified date, or, even worse, by the transformation of a continued gain into a loss. Nobody who is acquainted with the balance sheets of gambling banks would ever consider such a possibility. The lottery belongs, from this point of view, to the same class as roulette. Lotteries have been organized by certain governments for decades, and the results have always been in complete agreement with the assumption of constant values of the relative frequencies.

We thus see that the hypothesis of the existence of limiting values of the relative frequencies is well corroborated by a large mass of experience with actual games of chance. Only processes to which this hypothesis applies form the subject of our subsequent discussion.

THE PROBABILITY OF DEATH

The 'probability of death' is calculated by the insurance companies by a method very similar to the one which we have used to define the probability in the case of the game of dice. The first thing needed is to have an exact definition of the collective for each single case. As an example, we may mention the compiling of the German Life Tables Based on the Experience of Twenty-three Insurance Companies. These tables were calculated on the basis of 900,000 single observations on persons whose lives were insured with one of the twenty-three companies.[14] The observations covered the period from the moment of conclusion of the insurance contract until the cessation of this contract by death or otherwise. Let us consider, in particular, the following collective: 'All men insured before reaching the age of forty after complete medical examination and with the normal premium, the characteristic event being the death of the insured in his forty-first year.' Cases in which the occurrence or non-occurrence of this event could not be ascertained, e.g., because of a discontinuation of the insurance, were excluded from calculation. The number of cases which could be ascertained was 85,020. The corresponding number of deaths was 940. The relative frequency of deaths or the death-rate is therefore $940:85,020 = 0.01106$. This figure was accepted, after certain corrections which we do not need to be bothered with, as the probability of death occurring in the forty-first year for members of the above-described class of insured persons, i.e., for an exactly defined collective.

In this case 85,000 observations have been assumed to be sufficient for the relative frequency of deaths to become practically equal to its limiting value, that is, to a constant which refers to an indefinitely long series of observations of persons of the same category. This assumption is an arbitrary one and, strictly speaking, it would be wrong to expect that the above relative frequency agrees with the true probability to more than the first three decimal places. In other words, if we could increase the number of observations and keep calculating the relative frequency of deaths, we can only expect that the first three decimal places of the original death-rate, namely 0.011, will remain unchanged. All concerned in insurance business would prefer the death-rates to be calculated on a broader basis; this, however, is difficult for obvious practical reasons. On the other hand, no figure of this kind, however exact at the moment of its determination, can remain valid for ever. The same is true for all physical data. The scientists determine the acceleration due to gravity at a certain place on the surface of the earth, and continue to use this value until a new determination happens to reveal a change in it; local differences are treated in the same way. Similarly, insurance mathematicians are satisfied with the best data available at the moment, and continue to use a figure such as the above 0.011 until new and more accurate calculations become possible. In other words, the insurance companies continue to assume that out of 1000 newly insured men of the previously defined category, eleven will die in their forty-first year. No significance for any other category is claimed for this figure 0.011. It is utter nonsense to say, for instance, that Mr. X, now aged forty, has the probability 0.011 of dying in the course of the next year. If the analogous ratio is calculated for men and women together, the value obtained in this way is somewhat smaller than 0.011, and Mr. X belongs to this second collective as much as to that previously considered. He is, furthermore, a member of a great number of other collectives which can be easily defined, and for which the calculation of the probability of death may give as many different values. One might suggest that a correct value of the probability of death for Mr. X may be obtained by restricting the collective to which he belongs as far as possible, by taking into consideration more and more of his individual characteristics. There is, however, no end to this process, and if we go further and further into the selection of the members of the collective, we shall be left finally with this individual alone. Insurance companies nowadays apply the principle of so-called 'selection by insurance'; this means that they take into consideration the fact that persons who enter early into insurance contracts are on the average of a different type and have a different distribution of death ages from persons admitted to the insurance at a more advanced age. It is obviously possible to go further in this or other directions in the limitation of the collective. It is, however, equally obvious that in trying to take into account *all* the properties of an individual, we shall finally arrive at the stage of finding no other members of the collective at all, and the collective will cease to exist altogether.

FIRST THE COLLECTIVE—THEN THE PROBABILITY

I should like to dwell a little on this last point, which implies a characteristic difference between the definition of probability assumed in these lectures and that which has been generally accepted before. I have already stated this once in the following short sentence: 'We shall not speak of probability until a collective has been defined'. In this connexion, it is of interest to consider the diametrically opposite viewpoint expressed by one of the older authors, Johannes von Kries,[15] in a once widely read book on the principles of the theory of probability. He declares: '. . . I shall assume therefore a definite probability of the death of Caius, Sempronius or Titus in the course of the next year. If, on the other hand, the question is raised of the probability of a general event, including an indefinite number of individual cases—for instance, of the probability of a man of 40 living another 20 years, this clearly means the use of the word "probability" in another and not quite proper way—as a kind of abbreviation. If an expression of this kind should have any connexion with the true meaning of probability at all, this connexion may only consist in a comprehensive description of a certain number of single probabilities.'

My opinion is that the 'improper' use of the probability notion, as defined by von Kries, is in fact the only one admissible in the calculus of probability. This has been demonstrated in the foregoing paragraph by means of the same example of the death probability as was used by von Kries, and I have tried to show that any other conception is impossible. I consider, quite generally, the introduction of the expression 'probability in a collective' as an important 'improvement in word usage'. Two examples may help to elucidate this point further.

Consider a lottery with one million tickets. Imagine that the first prize has fallen to ticket No. 400,000. People will consider this an amazing and rare event; newspapers will discuss it, and everybody will think that this was a very improbable occurrence. On the other hand, the essence of a lottery is that all precautions have been taken to ensure the same probability for a win for all tickets, and No. 400,000 has therefore exactly the same chance of winning as all the other numbers, for instance No. 786,331—namely the probability 1/1,000,000. What shall we think about this paradox? Another example is given by Laplace[16] in his famous *Essai Philosophique*: In playing with small cards, on each of which is written a single letter, selecting at random fourteen of them and arranging them in a row, one would be extremely amazed to see the word 'Constantinople' formed. However, in this case again, the mechanism of the play is such as to ensure the same probability for each of the 26^{14} possible combinations of fourteen letters (out of the twenty-six letters of the alphabet). Why do we nevertheless assume the appearance of the word 'Constantinople' to be something utterly improbable?

The solution of these two seeming paradoxes is the same. The event that the first prize will fall to ticket No. 400,000 has, in itself, no 'probability' at all. A collective has to be defined before the word probability acquires a definite meaning. We may define this collective to consist of repeated draws of a lottery, the attribute of the particular draw being the number of the ticket drawn. In this collective each number has exactly the same probability as No. 400,000. However, in speaking of the 'improbability' of the number 400,000, we have in mind a collective of a different kind. The above-mentioned impression of improbability would be created not only by drawing the number 400,000, but all numbers of the same kind: 100,000, 200,000, etc. The collective with which we have to deal has

19

therefore only the two following attributes—either the number does end with five 0's, or it does not. The first-named attribute has the probability 0.00001, the second 0.99999, i.e., nearly 100,000 times larger. In an alternative between the draw of a number containing five 0's and that of a number not having this property, the second result has indeed a very much larger probability.

Exactly the same considerations apply to the second example. What astonishes us in the case of the word 'Constantinople' is the fact that fourteen letters, taken and ordered at random, should form a well-known word instead of unreadable gibberish. Among the immense number of combinations of fourteen letters (26^{14}, or about 10^{20}), not more than a few thousand correspond to words. The elements of the collective are in this case all the possible combinations of fourteen letters with the alternative attributes 'coherent' or 'meaningless'. The second attribute ('meaningless') has, in this collective, a very much larger probability than the first one, and that is why we call the appearance of the word 'Constantinople'—or of any other word—a highly improbable event.

In many appropriate uses of the probability notion in practical life the collective can be easily constructed. In cases where this proves to be impossible, the use of the word probability, from the point of view of the rational theory of probability, is an illegitimate one, and numerical determination of the probability value is therefore impossible. In many cases the collective can be defined in several ways and these are cases in which the magnitude of the probability may become a subject of controversy. It is only the notion of *probability in a given collective* which is unambiguous.

PROBABILITY IN THE GAS THEORY

We now return to our preliminary survey of fields in which the theory of probability can be applied, and consider the third example —that of molecular physics—rather more closely. In the investigation of the behaviour of molecules in a gas we encounter conditions not essentially different from those prevailing in the two applications of probability we have previously discussed. In this case, the collective can be formed, for instance, by all molecules present in the volume of gas enclosed by the walls of a cylinder and a piston. As attributes of the single elements (molecules), we may consider, for instance, the three rectangular components of their velocities, or the velocity vector itself. It is true that nobody has yet tried to measure the actual velocities of all the single molecules in a gas, and to calculate in this way the relative frequencies with which the different values occur. Instead, the physicist makes certain theoretical assumptions concerning these frequencies (or, more exactly, their limiting values), and tests experimentally certain consequences, derived on the basis of these assumptions. Although the possibility of a direct determination of the probability does not exist in this case, there is nevertheless no fundamental difference between it and the other two examples treated. The main point is that in this case, too, all considerations are based on the existence of constant limiting values of relative frequencies which are unaffected by a further increase in the number of elements concerned, i.e., by an increase in the volume of gas under consideration.

In order to explain the relation between this problem and the previous example of the probability of death established by direct counting, we may think of the following analogy. A surveyor may have to make calculations relating to a right-angled triangle, e.g., the evaluation of its hypotenuse by means of the Pythagorean theorem. His first step may be to establish by direct measurement that the angle in the triangle is sufficiently near to 90°. Another method which he can apply is to assume that this angle *is* 90°, to draw conclusions from this assumption, and to verify them by comparison with the experimental results. This is the situation in which the physicist finds himself when he applies statistical methods to molecules or other particles of the same kind. The physicists often say that the velocity of a molecule is 'in principle' a measurable quantity, although it is not possible to carry out this measurement in practice by means of the usual measuring devices. (At this stage we do not consider the modern development of the question of the measurability of molecular quantities.) Similarly, we can say that the relative frequency and its limiting value, the probability, are determined in the molecular collective 'in principle' in the same way as in the cases of the games of chance and of social statistics which we have previously discussed.

AN HISTORICAL REMARK

The way in which the probability concept has been developed in the preceding paragraphs is widely different from the one which the older textbooks of probability calculus used in formally defining their subject. On the other hand, our foundation of probability is in no contradiction whatsoever to the actual content of the probability concept used by these authors. In this sense, the first pages of Poisson's[17] famous textbook, *On the probability of the judgments of*

21

courts of justice, are very instructive. Poisson says that a certain phenomenon has been found to occur in many different fields of experience, namely, the fact which we have described above as the stabilization of relative frequencies with the increase in the number of observations. In this connexion, Poisson uses an expression which I have avoided up till now on account of a prevailing confusion regarding its interpretation. Poisson calls the fact that the relative frequencies become constant, after the sequence of experiments has been sufficiently extended, the Law of Large Numbers. He considers this law to be the basis of a theory of probability, and we fully agree with him on this point. In the actual investigations which follow the introduction, however, Poisson starts not from this law, but from the formal definition of probability introduced by Laplace.[18] (We shall have to speak about this definition later.) From it he deduces, by analytical methods, a mathematical proposition which he also calls the Law of Large Numbers. We shall see later on that this mathematical proposition means something very different from the general empirical rule called by the same name at the beginning of Poisson's book. This double use of the same expression to describe two widely different things has caused much confusion, and we shall have to return to this point again: it will form the subject of the fourth chapter. At that point, I shall also quote the exact words in which Poisson states the empirical rule of the constancy of the relative frequencies with large numbers of observations as the foundation of the theory of probability. In the meantime I ask you not to associate any definite meaning with the expression 'The Law of Large Numbers'.

Let me add that our conception of the *sequence of observations* as the cornerstone in the foundation of the theory of probability, and our definition of probability as the relative frequency with which certain events or properties recur in these sequences, is not something absolutely new. In a more dialectical form and without the immediate intention of developing a theory of probability calculus on this basis, the same ideas were presented as early as 1866 by John Venn[19] in his book *Logic of Chance*. The development of the so-called theory of finite populations by Theodor Fechner[20] and Heinrich Bruns[21] is closely related to our frequency theory of probability. Georg Helm,[22] who played a certain part in the foundation of the energy principle, expressed ideas very similar to ours in his paper on 'Probability Theory as the Theory of the Concept of Collectives', which appeared in 1902. These attempts, as well as many others which time does not allow us to enumerate, did not lead, and could not lead, to a complete theory of probability, because they failed to realize one decisive feature of a collective which we shall discuss in the following paragraph.

The condition that the relative frequencies of attributes should have constant limiting values is not the only one we have to stipulate when dealing with collectives, i.e., with sequences of single observations, mass phenomena, or repetitive events which may appropriately serve as a basis for the application of probability theory. Examples can easily be found where the relative frequencies converge towards definite limiting values, and where it is nevertheless not appropriate to speak of probability. Imagine, for instance, a road along which milestones are placed, large ones for whole miles and smaller ones for tenths of a mile. If we walk long enough along this road, calculating the relative frequencies of large stones, the value found in this way will lie around 1/10. The value will be exactly 0.1 whenever in each mile we are in that interval between two small milestones which corresponds to the one in which we started. The deviations from the value 0.1 will become smaller and smaller as the number of stones passed increases; in other words, the relative frequency tends towards the limiting value 0.1. This result may induce us to speak of a certain 'probability of encountering a large stone'. Nothing that we have said so far prevents us from doing so. It is, however, worth while to inquire more closely into the obvious difference between the case of the milestones and the cases previously discussed. A point will emerge from this inquiry which will make it desirable to restrict the definition of a collective in such a way as to exclude the case of milestones and other cases of a similar nature. The sequence of observations of large or small stones differs essentially from the sequence of observations, for instance, of the results of a game of chance, in that the first sequence obeys an easily recognizable law. Exactly every tenth observation leads to the attribute 'large', all others to the attribute 'small'. After having just passed a large stone, we are in no doubt about the size of the next one; there is no chance of its being large. If, however, we have cast a double 6 with two dice, this fact in no way affects our chances of getting the same result in the next cast. Similarly, the death of an insured person during his forty-first year does not give the slightest indication of what will be the fate of another who is registered next to him in the books of the insurance company, regardless of how the company's list was prepared.

This difference between the two sequences of observations is actually observable. We shall, in future, consider only such sequences of events or observations, which satisfy the requirements of complete lawlessness or 'randomness' and refer to them as collectives. In certain cases, such as the one mentioned above, where there is no collective properly speaking, it may sometimes be useful to have a short expression for the limiting value of the relative frequency. We shall then speak of the 'chance' of an attribute's occurring in an unlimited sequence of observations, which may be called an improper collective. The term 'probability' will be reserved for the limiting value of the relative frequency in a true collective which satisfies the condition of randomness. The only question is how to describe this condition exactly enough to be able to give a sufficiently precise definition of a collective.

DEFINITION OF RANDOMNESS: PLACE SELECTION

On the basis of all that has been said, an appropriate definition of randomness can be found without much difficulty. The essential difference between the sequence of the results obtained by casting dice and the regular sequence of large and small milestones consists in the possibility of devising a method of *selecting the elements* so as to produce a fundamental change in the relative frequencies.

We begin, for instance, with a large stone, and register only every second stone passed. The relation of the relative frequencies of the small and large stones will now converge towards 1/5 instead of 1/10. (We miss none of the large stones, but we do miss every second of the small ones.) If the same method, or any other, simple or complicated, method of selection is applied to the sequence of dice casts, the effect will always be nil; the relative frequency of the double 6, for instance, will remain, in all selected partial sequences, the same as in the original one (assuming, of course, that the selected sequences are long enough to show an approach to the limiting value). This impossibility of affecting the chances of a game by a system of selection, this uselessness of all systems of gambling, is the characteristic and decisive property common to all sequences of observations or mass phenomena which form the proper subject of probability calculus. In this way we arrive at the following definition: A collective appropriate for the application of the theory of probability must fulfil two conditions. First, the relative frequencies of the attributes must possess limiting values. Second, these limiting values must remain the same in all partial sequences which may be selected from the original one in an arbitrary way. Of course, only such partial sequences can be taken into consideration as can be extended indefinitely, in the same way as the original sequence itself. Examples of this kind are, for instance, the partial sequences formed by all odd members of the original sequence, or by all members for which the place number in the sequence is the square of an integer, or a prime number, or a number selected according to some other rule, whatever it may be. The only essential condition is that the question whether or not a certain member of the original sequence belongs to the selected partial sequence should be settled *independently of the result* of the corresponding observation, i.e., before anything is known about this result. We shall call a selection of this kind a *place selection*. The limiting values of the relative frequencies in a collective must be independent of all possible place selections. By place selection we mean the selection of a partial sequence in such a way that we decide whether an element should or should not be included without making use of the attribute of the element, i.e., the result of our game of chance.

THE PRINCIPLE OF THE IMPOSSIBILITY OF A GAMBLING SYSTEM

We may now ask a question similar to one we have previously asked: 'How do we know that collectives satisfying this new and more rigid requirement really exist?' Here again we may point to experimental results, and these are numerous enough. Everybody who has been to Monte Carlo, or who has read descriptions of a gambling bank, knows how many 'absolutely safe' gambling systems, sometimes of an enormously complicated character, have been invented and tried out by gamblers; and new systems are still being suggested every day. The authors of such systems have all, sooner or later, had the sad experience of finding out that no system is able to improve their chances of winning in the long run, i.e., to affect the relative frequencies with which different colours or numbers appear in a sequence selected from the total sequence of the game. This experience forms the experimental basis of our definition of probability.

An analogy presents itself at this point which I shall briefly discuss. The system fanatics of Monte Carlo show an obvious likeness to another class of 'inventors' whose useless labour we have been accustomed to consider with a certain compassion, namely, the ancient and undying family of constructors of 'perpetual-motion' machines. This analogy, which is not only a psychological one, is worth closer consideration. Why does every educated man smile nowadays when he hears of a new attempt to construct a perpetual-motion machine? Because, he will answer, he knows from the law of the conservation of energy that such a machine is impossible. However, the law of conservation of energy is nothing but a broad generalization—however firmly rooted in various branches of physics —of fundamental empirical results. The failure of all the innumerable attempts to build such a machine plays a decisive role among these. In theoretical physics, the energy principle and its various applications are often referred to as 'the principle of the impossibility of perpetual motion'. There can be no question of proving the law of conservation of energy—if we mean by 'proof' something more than the simple fact of an agreement between a principle and all the experimental results so far obtained. The character of being nearly self-evident, which this principle has acquired for us, is only due to the enormous accumulation of empirical data which confirm it. Apart from the unsuccessful attempts to construct a perpetual-motion machine—the interest of which is now purely historical—all the technical methods of transformation of energy are evidence for the validity of the energy principle.

By generalizing the experience of the gambling banks, deducing from it the Principle of the Impossibility of a Gambling System, and including this principle in the foundation of the theory of probability, we proceed in the same way as did the physicists in the case of the energy principle. In our case also, the naïve attempts of the hunters of fortune are supplemented by more solid experience, especially that of the insurance companies and similar bodies. The results obtained by them can be stated as follows. The whole financial basis of insurance would be questionable if it were possible to change the relative frequency of the occurrence of the insurance cases (deaths, etc.) by excluding, for example, every tenth one of the insured persons, or by some other selection principle. The principle of the impossibility of a gambling system has the same importance for the insurance companies as the principle of the conservation of energy for the electric power station: it is the rock on which all the calculations rest. We can characterize these two principles, as well as all far-reaching laws of nature, by saying that they are restrictions which we impose on the basis of our previous experience, upon our expectation of the further course of natural events. (This formulation goes back to E. Mach.) The fact that predictions of this kind have been repeatedly verified by experience entitles us to assume the existence of mass phenomena or repetitive events to which the principle of the impossibility of a gambling system actually applies. Only phenomena of this kind will be the subject of our further discussion.

In order to illustrate the randomness in a collective, I will show a simple experiment. It is again taken from the field of games of chance; this is only because experiments on subjects belonging to other fields in which the theory of probability finds its application require apparatus much too elaborate to be shown here.

I have a bag containing ninety round discs, bearing the numbers 1 to 90. I extract one disc from the bag at random, I note whether the number it bears is an odd or an even one and replace the disc. I repeat the experiment 100 times and denote all the odd numbers by 1's, and all even numbers by 0's. The following table shows the result:

1	1	0	0	0	1	1	1	0	1
0	0	1	1	0	0	1	1	1	1
0	1	0	1	0	0	1	0	0	0
0	1	0	0	1	0	0	1	1	1
0	0	1	1	0	0	0	0	1	1
0	1	1	1	1	0	1	0	1	0
1	0	1	1	1	1	0	0	1	1
0	0	1	1	0	1	1	1	0	1
0	0	1	1	0	0	1	1	0	1
0	1	1	0	0	0	1	0	0	0

Among 100 experimental results we find fifty-one ones; in other words, the relative frequency of the result 1 is 51/100. If we consider only the first, third, fifth draw, and so forth, i.e., if we take only the figures in the odd columns of the table, we find that ones appear in twenty-four cases out of fifty; the relative frequency is 48/100. Using only the numbers in the odd horizontal rows of the table, we obtain, for the relative frequency of the result 1, the value 50/100. We may further consider only those results whose place in the sequence corresponds to one of the prime numbers, i.e., 1, 2, 3, 5, 7, 11, 13, 17, 19, 23, 29, 31, 37, 41, 43, 47, 53, 59, 61, 67, 71, 73, 79, 83, 89 and 97. These twenty-six draws have produced thirteen 1's, the relative frequency is thus again exactly 50/100. Finally, we may consider only the 51 draws following a result 1. (A 'system' gambler might prefer to bet on 0 after 1 has just come out.) We find in this selection of results twenty-seven 1's, i.e., the relative frequency 27/51, or about 53/100. These calculations show that, in all the different selections which we have tried out, the 1's always appear with a relative frequency of about 1/2. I trust that this conveys the feeling that more extensive experiments, which I am not able to carry out here because of the lack of time, would demonstrate the phenomenon of randomness still more strikingly.

It is of course possible, after knowing the results of the hundred draws, to indicate a method of selection which would produce only 1's, or only 0's, or 1's and 0's in any desired proportion. It is also possible that in some other group of a hundred experiments, analogous to the one just performed, one kind of selection may give a result widely different from 1/2. The principle of randomness requires only that the relative frequency should converge to 1/2 when the number of results in an arbitrarily selected partial sequence becomes larger and larger.

SUMMARY OF THE DEFINITION

I do not need to insist here on mathematical details and such considerations which are only necessary for making the definitions complete from a mathematical point of view. Those who are interested in this question may refer to my first publication on the foundation of probability calculus (1919) or to my textbook of the theory of probability (1931), presenting the theory in a simplified and, it seems to me, improved form. (See Notes and Addenda, p. 224.) In my third lecture I will deal with various basic questions and with different objections to my definition of probability and these I hope to be able to refute. I trust that this discussion will dispel those doubts which may have arisen in your minds and further clarify certain points.

In closing this first lecture, may I summarize briefly the propositions which we have found and which will serve as a basis for all future discussions. These propositions are equivalent to a definition of mathematical probability, in the only sense in which we intend to use this concept.

1. It is possible to speak about probabilities only in reference to a properly defined collective.

2. A collective is a mass phenomenon or an unlimited sequence of observations fulfilling the following two conditions: (i) the relative frequencies of particular attributes within the collective tend to fixed

limits; (ii) these fixed limits are not affected by any place selection. That is to say, if we calculate the relative frequency of some attribute not in the original sequence, but in a partial set, selected according to some fixed rule, then we require that the relative frequency so calculated should tend to the same limit as it does in the original set.

3. The fulfilment of the condition (ii) will be described as the Principle of Randomness or the Principle of the Impossibility of a Gambling System.

4. The limiting value of the relative frequency of a given attribute, assumed to be independent of any place selection, will be called 'the probability of that attribute within the given collective'. Whenever this qualification of the word 'probability' is omitted, this omission should be considered as an abbreviation and the necessity for reference to some collective must be strictly kept in mind.

5. If a sequence of observations fulfills only the first condition (existence of limits of the relative frequencies), but not the second one, then such a limiting value will be called the 'chance' of the occurrence of the particular attribute rather than its 'probability'.

The Elements of the Theory of Probability

IN the first lecture I have already mentioned that the conception which I developed there, defining probability as the limiting value of an observable relative frequency, has its opponents. In the third lecture I intend to examine the objections raised against this definition in greater detail. Before doing this, I should like, however, to describe briefly the application of the fundamental definitions to real events, how they can be used for solving practical problems; in short, I shall discuss their general value and utility. The applicability of a theory to reality is, in my opinion, if not the only, then the most important test of its value.

THE THEORY OF PROBABILITY IS A SCIENCE SIMILAR TO OTHERS

I begin with a statement which will meet with the immediate opposition of all who think that the theory of probability is a science fundamentally different from all the other sciences and governed by a special kind of logic. It has been asserted—and this is no overstatement—that whereas other sciences draw their conclusions from what we know, the science of probability derives its most important results from what we do not know. 'Our absolute lack of knowledge concerning the conditions under which a die falls,' says Czuber,[1] 'causes us to conclude that each side of the die has the probability 1/6.' If, however, our lack of knowledge were as complete as Czuber assumes it to be, how could we distinguish between the two pairs of dice shown in the preceding lecture? Yet, the probability of casting '6' with one of them is considerably different from 1/6—at least, according to *our* definition of probability.

In fact, we will have nothing to do with assumptions as fantastic as

30

that of a distinct kind of logic used in the theory of probability. Twice two are four; B and the contrary of B cannot both follow from one and the same true premise—any more in the theory of probability than elsewhere. And *ex nihilo nihil* is true in the theory of probability as well. Like all the other natural sciences, the theory of probability starts from observations, orders them, classifies them, derives from them certain basic concepts and laws and, finally, by means of the usual and universally applicable logic, draws conclusions which can be tested by comparison with experimental results. In other words, in our view the theory of probability is a normal science, distinguished by a special subject and not by a special method of reasoning.

THE PURPOSE OF THE THEORY OF PROBABILITY

From this sober scientific point of view, which assumes that the same laws of reasoning, and the same fundamental methods are applicable in the theory of probability as in all other sciences, we can describe the purpose of the theory of probability as follows: Certain collectives exist which are in some way linked with each other, e.g., the throwing of one or the other of two dice separately and the throwing of the same two dice together form three collectives of this kind. The first two collectives determine the third one, i.e., the one where both dice are thrown together. This is true so long as the two dice, in falling together, do not influence each other in any way. If there is no such interaction, experience has shown that the two dice thrown together give again a collective such that its probabilities can be derived, in a simple way, from the probabilities in the first two collectives. This derivation and nothing else is here the task of probability calculus. In this problem the given quantities are the six probabilities of the six possible results of casting the first die and the six similar probabilities for the second die. A quantity which can be calculated is, for example, the probability of casting the sum '10' with the two dice.

This is very much like the geometrical problem of calculating the length of a side of a triangle from the known lengths of the two other sides and the angle formed by them. The geometer does not ask how the lengths of the two sides and the magnitude of the angle have been measured; the source from which these initial elements of the problem are taken lies outside the scope of the geometrical problem itself. This may be the business of the surveyor, who, in his turn, may have to use many geometrical considerations in his work. We shall find an exact analogy to these relations in the interdependence of statistics and probability. Geometry proper teaches us only how to determine certain unknown quantities from other quantities which are supposed to be known—quite independently of the actual values of these known quantities and of their derivation. The calculus of probability, correctly interpreted, provides us, for instance, with a formula for the calculation of the probability of casting the sum '10', or the 'double 6', with two dice, a formula which is of general validity, whichever pair of dice may be used, e.g., one of the two pairs discussed in the preceding lecture, or another pair formed from these four dice, or a completely new and different set. The six probabilities for the six sides of the first die, and the corresponding set of six probabilities for the second die, may have any conceivable values. The source from which these values are known is irrelevant, in the same way in which the source of knowledge of the geometrical data is irrelevant for the solution of the geometrical problem in which these data are used.

A great number of popular and more or less serious objections to the theory of probability disappear at once when we recognize that the exclusive purpose of this theory is to determine, from the given probabilities in a number of initial collectives, the probabilities in a new collective derived from the initial ones. A mathematician teased with the question, 'Can you calculate the probability that I shall not miss the next train?', must decline to answer it in the same way as he would decline to answer the question, 'Can you calculate the distance between these two mountain peaks?'—namely, by saying that a distance can only be calculated if other appropriate distances and angles are known, and that a probability can only be determined from the knowledge of other probabilities on which it depends.

Because certain elements of geometry have for a long time been included in the general course of education, every educated man is able to distinguish between the practical task of the land surveyor and the theoretical investigation of the geometer. The corresponding distinction between the theory of probability and statistics has yet to be recognized.

THE BEGINNING AND THE END OF EACH PROBLEM MUST BE PROBABILITIES

We all know, I think, that in each mathematical problem there are, on the one hand, a number of known quantities or data, and, on the other, certain quantities which are to be determined and which, after this determination, are called results. The conclusion at which we arrived in the last paragraph can be restated by saying: In a problem of probability calculus, the data as well as the results are probabilities. Emphasis was laid above on the first part of this statement, namely, the starting point of all probability calculations. I should like to add a few words concerning the second part.

The result of each calculation appertaining to the field of probability is always, as far as our theory goes, nothing else but a probability, or, using our general definition, the relative frequency of a certain event in a sufficiently long (theoretically, infinitely long) sequence of observations. The theory of probability can never lead to a definite statement concerning a single event. The only question that it can answer is: what is to be expected in the course of a very long sequence of observations? It is important to note that this statement remains valid also if the calculated probability has one of the two extreme values 1 or 0.

According to the classical theory of probability, and to some new versions of this theory, the probability value 1 means that the corresponding event will certainly take place. If we accept this, we are admitting, by implication, that the knowledge of a probability value can enable us, under certain circumstances, to predict with certainty the result of any one of an infinite number of experiments. If, however, we define probability as the limiting value of the relative frequency, the probability value 1 does not mean that the corresponding attribute must be found in every one of the elements forming the collective. This can be illustrated by the following example:

Imagine an infinite sequence of elements distinguished by the two different attributes A and B. Assume that the sequence has the following structure: First comes an A, then a B, then again an A, then a group of two consecutive B's, again one A, then a group of three B's, and so on, the single A's being separated by steadily growing groups of B's:

$$ABABBABBBBABBBBBABBBBB \ldots$$

This is a regular sequence of symbols; it can be represented by a mathematical formula, and it is easily ascertained that, with the increasing number of elements, the relative frequency of the attribute A converges towards 0, whereas the relative frequency of B converges towards unity. The same thing can occur also in irregular sequences. Once an attribute is rare enough, it is possible that its relative frequency, although never attaining the value 0, converges to this value with increasing length of the sequence. In other words, its limiting value may be 0. We see, therefore, that the probability 0 means only a very rare—we may say, an infinitely rare—occurrence of an event, but not its absolute impossibility. In the same way the probability 1 means that the corresponding attribute occurs nearly always, but not that it must be necessarily found in each observation. In this way the indeterminate character of all statements of the probability theory is maintained in the limiting cases as well.

It remains now to give a more detailed consideration as to what is meant by the *derivation of one collective from another*, an operation to which we have repeatedly referred. It is only from a clear conception of this process that we can hope to recognize fully the nature of the fundamental task of probability calculus. We shall begin this investigation by introducing a new, simple expression which will permit us to make our statements in a clearer and simpler way. It is at first only an abbreviation; later, however, it will lead us to a slight extension of the concept of a collective.

DISTRIBUTION IN A COLLECTIVE

The elements or members of a collective are distinguished by certain attributes, which may be numbers, as in the case of the game of dice, colours, as in roulette (*rouge et noir*), or any other observable properties. The smallest number of different attributes in a collective is two; in this case we call it a *simple alternative*. In such a collective there are only two probabilities, and, obviously, the sum of these two must be 1. The game of 'heads or tails' with a coin is an example of such an alternative, with the two distinctive attributes being the two different faces of the coin. Under normal conditions, each of these attributes has the same probability, 1/2. In problems of life insurance we also deal with a simple alternative. The two possible events are, for example, the death of the insured between the first and the last day of his forty-first year, and his survival beyond this time. In this example, the probability of the first event is 0.011 and that of the second one 0.989. In other cases, such as the game of dice, more than two attributes are involved. A cast with one die can give six different results, corresponding to the six sides of the die. There are six distinct probabilities and their sum is again 1. If all the six results are equally probable, the single probabilities all have the value 1/6. We call a die of this kind an unbiased one. However, the die may be biased; the six probabilities will still be proper fractions with the sum 1, although not all equal to 1/6. The values of these six probabilities must be known if the corresponding collective is to be considered as given.

34

It is useful to have a short expression for denoting the whole of the probabilities attached to the different attributes in a collective. We shall use for this purpose the word *distribution*. If we think of the distribution of chances in a game of chance, the reasons for this choice of term will be easily understood. If, for instance, six players bet, each on one of the six different sides of a die, the chances are 'distributed' in such a way that the relative chance of each of the players is equal to the probability of the side which he has chosen. If the die is an unbiased one, all the chances are equal; they are then uniformly 'distributed'. With a biased die, the distribution of chances is non-uniform. In the case of a simple alternative the whole distribution consists of two numbers only, whose sum is 1. To illustrate the meaning of the word 'distribution', one can also think of how the different possible attributes are distributed in the infinite sequence of elements forming the collective. If, for instance, the numbers 1/5, 3/5, and 1/5 represent the distribution in a collective with three attributes A, B, and C, the probabilities of A and C being 1/5 each, and that of B being 3/5, then in a sufficiently long sequence of observations we shall find the attributes A, B, and C 'distributed' in such a way that the first and third of them occur in 1/5 of all observed cases and the second in the remaining 3/5.

PROBABILITY OF A HIT; CONTINUOUS DISTRIBUTION

The concept of distribution defined above leads to the consideration of certain cases which have been so far left aside. Imagine a man shooting repeatedly at a target. This is a repetitive event; there is nothing to prevent it from being, in principle, continued indefinitely. By assigning a number to each concentric part of the target, beginning with 1 for the bull's-eye, and ending with the space outside the last ring, we can characterize each shot by means of a number. So far there is nothing unfamiliar in the example; the number of different attributes of the collective which, together with the corresponding probabilities, makes up the distribution, is equal to the number of different concentric regions of the target. Naturally, in order to be able to speak of probabilities at all, we must assume that the usual conditions concerning the existence of limits of frequencies and of randomness are satisfied.

We may, however, treat the same example in a slightly different way. We may measure the distance of each hit from the centre of the target and consider this distance as the attribute of the shot in question, instead of the number assigned to the corresponding ring.

'Distance' is then also nothing but a number—the number of units of length which can be marked off on a straight line between two given points. As long as we measure this distance in whole centimetres only, the situation is not very different from that previously described, and each shot is characterized, as before, by an integer. If the radius of the target is 1 metre, the number of different possible attributes is 101, namely, the integers from 0 to 100; consequently there are 101 different probabilities, and the distribution consists of 101 proper fractions giving the sum 1. Everyone, however, feels that the measure of a distance in centimetre units is not an adequate expression of the notion of distance. There are more than just 101 different distances between 0 and 1 metre. Geometry teaches us that distance is a continuous variable, which may assume every possible value between 0 and 100, i.e., values belonging to the infinite set of fractional numbers as well as those belonging to the finite set of whole numbers in this interval. We arrive in this way at the idea of a collective with an infinite number of attributes. In such cases the classical books speak of *geometrical probabilities*, which are thus contrasted with *arithmetical* ones, where the number of attributes is finite. We do not propose to question here the appropriateness of these terms. However, in a case like the present one, in order to describe the distribution in the same way as before, one would need an infinite set of fractions to represent the corresponding probabilities, and to present such a set is obviously impossible. Fortunately, the way of solving difficulties of this kind was discovered long ago, and everybody who has some slight knowledge of analysis knows how to proceed in a case like this.

PROBABILITY DENSITY

To explain how it is possible to describe distributions in which an infinite continuum of attributes is involved, we may consider an analogous case in another field. Imagine that we have to distribute a certain mass, say 1 kg, along a straight line 1 metre long. As long as the number of loaded points remains finite, the distribution consists of a finite number of fractions—fractions of a kilogram assigned to any such point. If, however, the weight has to be distributed continuously along the whole length of the straight line, e.g., in the form of a rod of nonuniform thickness, 1 metre long and of 1 kg weight, we can no longer speak of single loaded points. Nevertheless, the meaning of the expression *distribution of mass* is quite clear in this case as well. For example, we say that more mass is concentrated in a certain element of length in the thicker part of the rod than in an equal element in its thinner part, or that the mass density (mass per unit length) is greater in the thicker and smaller in the thinner part. Again, if the rod is uniformly thick, we speak of a uniform distribution of mass. Generally speaking, the distribution is full described by the indication of the mass density at each point of the line.

It is easy to extend this concept to the case of hits on a target. To each segment of distance between 0 and 100 cm there corresponds a certain probability of finding a hit in it, and the distribution of these hits is described by their density (number of hits per unit length) in each part of the target. We take the liberty of introducing a new expression, 'probability density',[2] and state: If a collective contains only a finite number of attributes, with no continuous transition between them, then its distribution consists of a finite number of probabilities corresponding to these attributes. If, however, the attributes are continuously variable quantities, e.g., distances from a fixed point, the distribution is described by a function representing the probability density per unit of length over the range of the continuous variable.

Let us take again the case of shots at a target. Assuming that the shots were fired blindly, we may expect the number of shots hitting a ring near the outer circumference of the target to be greater than that of the shots hitting a ring nearer to the centre, because the surface of the former is larger. We may thus expect the probability density to increase proportionally to the radius (or to some power of the radius).

We shall have to deal later with certain problems arising from the existence of collectives with continuously varying attributes; we shall further discuss the generalization of this concept so as to include attributes that are continuous in more than one dimension, i.e., densities on surfaces or volumes rather than on lines. At this stage of our discussion we merely mention these questions in order to illustrate the notion of distribution, and to indicate its wide range. By using this concept, we can now give a more precise formulation of the purpose of the theory of probability.

From one or more well-defined collectives, a new collective is derived (by methods which will be described in the following paragraphs). The purpose of the theory of probability is to calculate the distribution in the new collective from the known distribution (or distributions) in the initial ones.

THE FOUR FUNDAMENTAL OPERATIONS

The above statement contains the concept of a 'collective derived from others' and this requires closer consideration. How can a new collective be derived from a given one? Unless the manner in which such a derivation is made is clearly explained, all that has been said so far is in danger of becoming devoid of meaning. There are four, and only four, ways of deriving a collective and all problems treated by the theory of probability can be reduced to a combination of these four fundamental methods. Most practical problems involve the application, often a repeated one, of several fundamental operations. We shall now consider each of them in turn. In each of the four cases, our basic task is to compute the new distribution in terms of the given ones, and I do not expect that this will give rise to any great difficulty. The first two of the four fundamental operations are of surprising simplicity. Some of you may even think that I am trying to avoid the real mathematical difficulties of the theory of probability, difficulties which you think are bound to exist on account of the large number of formulæ usually found in textbooks. This is far from my purpose. By saying that all the operations by which different collectives are brought into mutual relation in the theory of probability can be reduced to four relatively simple and easily explained types, I do not suggest that there are no difficulties in the solution of problems which we may encounter. Such difficulties arise from the complicated combination of a great number of the four fundamental operations. Remember that algebra, with all its deep and intricate problems, is nothing but a development of the four fundamental operations of arithmetic. Everyone who understands the meaning of addition, subtraction, multiplication, and division holds the key to all algebraic problems. But the correct use of this key requires long training and great mental effort. The same conditions are found in the calculus of probability. I do not plan to teach you in a few lectures how to solve problems which have occupied the minds of a Bernoulli or a Laplace, as well as of many great mathematicians of our time. On the other hand, nobody would willingly give up the knowledge of the four fundamental operations of arithmetic, even if he were free of all mathematical ambition and had no need to perform any mathematical work. This knowledge is valuable, not only from the point of view of practical utility, but also for its educational value. By explaining briefly the four fundamental operations of the theory of probability, I hope to achieve the same two objects: to give you tools for solving occasionally a simple probability problem, and, what is more important, to give you some understanding of what the theory of probability means; this is a matter of interest to every educated person.

FIRST FUNDAMENTAL OPERATION: SELECTION

The first of the four fundamental operations by which a new collective can be derived from one (or several) initial ones is called *selection*. Imagine, for instance, a collective consisting of all casts made with a certain die, or of all games on a certain roulette table. New collectives can be formed, for instance, by selecting the first, fourth, seventh . . . casts of the die, or the second, fourth, eighth, sixteenth . . . games of roulette—generally speaking, by the selection of elements occupying certain places in the total sequence of the original collective. The attributes in the new collective remain the same as in the original one, namely, the number of points on the die, or the colours 'red' or 'black' in roulette. We are interested in the probabilities in the new collective, e.g., the probabilities of 'red' and 'black' in a selection of roulette games consisting only of games the order-numbers of which are, say, powers of 2 in the original sequence. According to an earlier statement concerning the properties of collectives, especially their randomness, the answer to this question is obvious: the probabilities remain unchanged by the transition from the initial collective to the new one formed by selection. The six probabilities of the numbers 1 to 6 are the same in the selected sequence of games of dice as they were in the original one. This, and nothing more, is the meaning of the condition of randomness imposed on all collectives. The whole operation is so simple that it hardly requires any further explanation. We therefore proceed immediately to the following exact formulation:

From a given collective, many new ones can be formed by selections of various kinds. The selected collective is a partial sequence derived from the complete sequence by the operation of *place selection*. The attributes in the selected collective are the same as in the original one. The distribution within the new collective is the same as in the original one.

SECOND FUNDAMENTAL OPERATION: MIXING

The second method of formation of a new collective from a given one is scarcely more complicated: this is the operation called *mixing*.

First, let us consider an example: take the same game of dice as in the previous example, the elements of the collective being the consecutive casts, and the attributes, the six different possible results. The following question can be asked: What is the probability of casting an even number? The answer is well known. Of the six numbers 1 to 6, three are even ones, 2, 4, and 6. The probability of an even number is the sum of the probabilities of these three results. I hardly expect anyone to doubt the correctness of this solution; it is also easy to deduce it by means of the definition of probability as the limiting value of relative frequency. The general principle underlying the operation is easily recognizable. We have constructed a new collective consisting of the same elements as before, but with new attributes. Instead of the six former attributes 1 to 6, we have now two new ones, 'even' and 'odd'. The essential point is that several original attributes are covered by a single new one. It would be different if an original attribute were replaced by several new ones, for this would make the calculation of the new probabilities from the initial ones impossible. The term 'mixing' is chosen to connote that several original attributes are now mixed together to form a single new attribute. We can also say that mixing is performed on several elements differing originally by their attributes, but now forming a unit in the new collective.

INEXACT STATEMENT OF THE ADDITION RULE

Perhaps some of you will remember from school days references to the probability of 'either-or' and the following proposition concerning the calculation of unknown probabilities from known ones: The probability of casting either 2 or 4 or 6 is equal to the sum of the probabilities of each of these results separately. This statement is, however, inexact; it remains incomplete even if we say that only probabilities of mutually exclusive events can be added in this way. The probability of dying in the interval between one's fortieth and one's forty-first birthday is, say, 0.011, and that of marrying between the forty-first and the forty-second birthdays 0.009. The two events are mutually exclusive; nevertheless we cannot say that a man entering his forty-first year has the chance $0.011 + 0.009 = 0.020$ of either dying in the course of this year or marrying in the course of the following year.

The clarification and the ensuing correct formulation of the mixing operation can only be achieved by having recourse to the concept of the collective. The difference between the correct formulation of the addition rule and the incorrect one follows from the principle that only such probabilities can be added as are attached to different attributes *in one and the same collective*. The operation consists in mixing only attributes of this kind. In the above example, however, the two probabilities belonged to two different collectives. The first collective was that of all men aged forty, and the two attributes were the occurrence and nonoccurrence of death in the course of the forty-first year of age. The second collective was formed of men who have attained their forty-first year, and who were divided into groups characterized by the occurrence or nonoccurrence of the event of marriage in the course of the following year. Both collectives are simple alternatives. The only possible mixing operation in each of them is the addition of the two probabilities, of life and death, or of marrying and remaining single—giving in each case the sum 1. It is not permissible to mix together attributes belonging to two different collectives.

Another example which shows clearly the insufficiency of the usual 'either-or' proposition follows: Consider a good tennis player. He may have 80% probability of winning in a certain tournament in London. His chance of winning another tournament in New York, beginning on the same day, may be 70%. The possibility of playing in both tournaments is ruled out, hence, the events are mutually exclusive, but it is obviously nonsense to say that the probability of his winning either in London or in New York is $0.80 + 0.70 = 1.50$. In this case again, the explanation of the paradox lies in the fact that the two probabilities refer to two different collectives, whereas the addition of probabilities is only allowed within a single collective.

UNIFORM DISTRIBUTION

A very special case of mixing which occurs often is sometimes given the first place in the presentation of the theory of probability; it is even assumed that it forms the basis of every calculation of probabilities. We have previously asked: What is the probability of casting an even number of points with a die? The general solution of this problem does not depend on the special values of the probabilities involved, i.e., those of the results, 2, 4, 6. The die may be an unbiased one, with all six probabilities equal to 1/6; the sum is in this case $1/6 + 1/6 + 1/6 = 1/2$. The die may, however, be one of the biased ones, such as we have already used several times, and the six probabilities may be different from 1/6. The rule, according to which the probability of an even number of points is equal to the sum of the probabilities of the three possible even numbers, remains valid in either case. The special case which we are now going to consider is that first mentioned, i.e., that of the unbiased die, or, as we are going to call it, the case of the 'uniform' distribution of probabilities.

In this case a correct result may be obtained in a way slightly different from the one we have used before. We begin with the fact that six different results of a cast are possible and each of them is equally likely. We now use a slightly modified method of reasoning: we point out that, out of the six possible results, three are 'favourable' to our purpose (which is to cast an even number) and three are 'unfavourable'. The probability of an even number, that is $3/6 = 1/2$, is equal to the ratio of the number of favourable results to the total number of possible results. This is obviously a special case of a general rule, applicable to cases in which all attributes in the initial collective have equal probabilities. We may assume, for instance, that the number of possible attributes in n, and that the probability of the occurrence of each of them is $1/n$. Assuming further that m among the n attributes are mixed together to form a new one, we find, by means of the addition rule, that the probability of the new attribute (in the new collective) is a sum of m terms, each equal to $1/n$. In other words the probability is m/n, or equal to the ratio of the number of favourable attributes to the total number of different original attributes. Later, we shall show how this rule has been misused to serve as a basis for an apparent definition of probability. For the time being we shall be satisfied with having clearly stated that the determination of probabilities by counting the number of equally probable, favourable and unfavourable, cases is merely a very special case of the derivation by mixing of a new collective from one initially given.

I have already made use of this special form of the mixing rule in my first lecture, without explicit mention. We spoke there of two collectives, whose elements were the consecutive draws in a lottery. In the first case, the attributes considered were all the different numbers of the lottery tickets; in the second case, the numbers ending with five 0's were bracketed together. Ten numbers ending with five 0's exist between 0 and one million. By adding their probabilities, with the assumption of a uniform distribution of probabilities in the initial collective, we found the probability of drawing a number ending with five 0's to be equal to 10 in a million, or 0.00001.

SUMMARY OF THE MIXING RULE

I will now briefly formulate the mixing rule, as derived from the concept of a collective.

Starting with an initial collective possessing more than two attributes, many different new collectives can be derived by 'mixing'; the elements of the new collective are the same as those of the old one, their attributes are 'mixtures' of those of the initial collective, e.g., all odd numbers or all even numbers, rather than the individual numbers, 1, 2, 3, . . . The distribution in the new collective is obtained from the given distribution in the initial collective by summing the probabilities of all those original attributes which are mixed together to form a single attribute in the new collective.

The practical application of this rule has already been illustrated by simple examples. I would mention in passing that this rule can be extended to include collectives, of the kind previously discussed, that have a continuous range of attributes. Higher mathematics teaches us that, in a case of this kind, addition of probabilities is replaced by an operation called integration, which is analogous to addition but less easily explained. Remember, for instance, the example of shooting at a target. Let us assume that the probability density is known for all distances from the centre of the target. The probability of a hit somewhere between 0.5 m and 1 m from the centre, i.e., in the outer half of the target, can be calculated by a mixing operation, involving the integration of the density function between the limits 0.5 m to 1.0 m^3. These indications are sufficient for those who are familiar with the fundamental concepts of analysis. Others may be sure that, although these generalizations are undoubtedly necessary for the solution of many problems, they are irrelevant from the point of view of those general principles which are our only concern in these lectures.

THIRD FUNDAMENTAL OPERATION: PARTITION

After having considered the first two operations by which a new collective can be derived, those of selection and of mixing, we now turn to the third one, which I call *partition*. The choice of this term will soon be clear to you; the word suggests a certain analogy to the arithmetical term 'division', the operation in question being in fact a 'division of probabilities'. To help you understand this third operation, I shall start with the same collective which served for the explanation of the first two operations, namely that formed by a series of throws of a die from a dice-box. The six attributes are again the numbers of points appearing on the top side of the die. The corresponding six probabilities have the sum 1, without being necessarily equal to 1/6 each. The new problem which we are going to discuss now and to solve by means of the operation which we call partition is the following: What is the probability that a result which we already know to be an even number will be equal to 2? This question may appear somewhat artificial, but it can easily be given a form which is often met in real life.

Imagine that you are standing at a bus stop where six different lines pass by. Three of them are served by double-decked buses and three by single-decked buses. The first ones may bear the numbers 2, 4, 6; the second ones, the numbers 1, 3, 5. When a bus approaches the stop, we recognize from afar to which of the two groups it belongs. Assuming that it has a double deck, what is the probability of its belonging to line No. 2? To solve this problem, we must of course know the six original probabilities (or, practically speaking, the relative frequencies) of the six services. Assuming that they are all equally frequent, and that the probabilities are therefore all equal to 1/6 (thus corresponding to the case of an unbiased die), the answer is easy: the probability of a double-decked bus being No. 2 is 1/3. One of the arguments by which we can arrive at this result is as follows: There are three different, equally probable possibilities; only one of them is a favourable one; its probability is, therefore, according to a rule quoted previously, equal to 1/3. This method of calculation is, however, not always applicable; it cannot be used if the six bus lines pass the stop with different frequencies, or if the six sides of the die have different probabilities. We arrive at a general solution and at a general statement of the problem by inquiring into the nature of the new derived collective. We are by now sufficiently accustomed to the idea that the expression 'probability of an event' has no exact meaning unless the collective in which this event is to be considered has been precisely defined.

For the sake of simplicity, let us return to the example of the die. The new derived collective may be described as follows. It is formed of elements of the initial collective but not of all its elements. In fact, it contains only those casts of the die which are distinguished by having the common attribute 'even number of dots'. The attributes within the new collective are the same as in the initial collective, namely, 'numbers of dots on the upper side of the die', but, whereas in the initial collective there were six different attributes 1, 2, . . ., 6, in the derived collective there are only three, 2, 4, and 6. We say that the new collective resulted from a partition into two categories of the elements of the original collective. One of them, the elements of which are distinguished by the common attribute 'even number', forms the derived collective. It is important to realize that this partition is something quite different from the place selection which we have discussed before. The latter consists in selecting certain elements out of the initial collective, according to a rule, ignoring the attributes while specifying the order numbers of the elements to be selected for the new collective. We have seen that the probabilities in a collective obtained in this way are the same as in the original one. On the other hand, when dealing with partition, the decision whether a given element is to be selected to take part in the derived collective is specifically based on what its attribute is. As a result, the probabilities within the derived collective are essentially different from those within the original one and the manner of their change is the subject of the following section.

PROBABILITIES AFTER PARTITION

Let us assume that the probabilities of the six possible results (1 to 6 points) in the original collective are 0.10, 0.20, 0.15, 0.25, 0.10, and 0.20 respectively, their sum being 1. It is unimportant whether we think of the game of dice, or the case of the buses. The way in which the six probabilities have been derived is equally irrelevant. We now add together the probabilities of all even numbers, i.e., the fractions 0.20, 0.25, and 0.20; this gives the sum 0.65 as the probability for the occurrence of any one of the even numbers (second fundamental problem, mixing, solved by the addition rule). According to our concept of probability, the frequency of 'even' results in a sufficiently long sequence of observations, is equal to 65 %. About 6500 elements among the first 10,000 observed have even numbers as their attributes. About 2000 of them have the attribute 2, since 0.20 is the frequency of this attribute. We are now going to form a new collective by excluding from the initial one all elements whose attributes are odd numbers. Among the first 6500 elements of the new collective, we find 2000 elements with the attribute 2; the relative frequency of this attribute is therefore 2000/6500 = 0.308. Since the calculation which we have just carried out is, strictly speaking, only valid for an infinitely long sequence of observations, the result which we have obtained, the fraction 0.308, represents already the limiting value of the relative frequency of the attribute 2 in the new collective; in other words, 0.308 is the probability of this attribute. The general rule for the solution of problems of this kind is now easily deduced from this special case. The first step is to form the sum of the given probabilities of all those attributes which are to be retained in the partition, i.e., the 2's, 4's, and 6's in our example. The next step is to divide by this sum the probability of the attribute about which we are inquiring (2 in the chosen example); 0.20/0.65 = 0.308. The procedure is in fact that of a division of probabilities.

INITIAL AND FINAL PROBABILITY OF AN ATTRIBUTE

It is useful to introduce distinct names for the two probabilities of the same attribute, the given probability in the initial collective and the calculated one in the new collective formed by partition. The current expressions for these two probabilities are not very satisfactory, although I cannot deny that they are impressive enough. The usual way is to call the probability in the initial collective the *a priori*, and that in the derived collective the *a posteriori* probability. The fact that these expressions suggest a connexion with a well-known philosophical terminology is their first deficiency in my eyes. Another one is that these same expressions, *a priori* and *a posteriori*, are used in the classical theory of probability in a different sense as well, namely, to distinguish between probabilities derived from empirical data and those assumed on the basis of some hypothesis; such a distinction is not pertinent in our theory. I prefer, therefore, to give to the two probabilities in question less pretentious names, which have less far-reaching and general associations. I will speak of *initial probability* and *final probability*, meaning by the first term the probability in the original collective, and by the second one, the probability (of the same attribute) in the collective derived by partition. In our numerical example the attribute 2 (two points on the die, or the bus line 2), has the initial probability 0.20, and the final probability 0.20/0.65 = 0.308. This means simply that this attribute has the probability 0.20 of being found among *all* the elements of the sequence, and the probability 0.308 of being found among those elements which resulted in an even number.

THE SO-CALLED PROBABILITY OF CAUSES

Another expression which I cannot leave unmentioned, and which I find equally misleading and confusing, is often used in connexion with the problem of partition. In the discussion of cases similar to that treated in the preceding paragraphs, i.e., the probability of the number 2 among the even numbers 2, 4, and 6, it is often argued that the appearance of an even number may have three different 'causes', or can be explained by three different 'hypotheses'. The possible 'causes' or 'hypotheses' are nothing else but the appearance of one of the three numbers 2, 4, 6. The above-calculated final probability 0.308 is correspondingly described as the probability of the appearance of an even number being 'caused' by the number 2. In this way an apparently quite new and special chapter of probability calculus is opened, dealing with the 'probability of causes' or the 'probability of hypotheses', instead of the usual 'probability of events'. The partition problem is usually presented in this theory in the following form:

Three urns, filled with black and white balls, are placed on the table. We consider an initial collective, each element of which is composed of two separate observations. The first observation consists in selecting at random one of the three urns and stating its number, 1, 2, or 3. The second observation consists in drawing a ball out of the urn previously selected and in noting its colour. Thus the attribute of each single element of the initial collective consists of the colour of the ball drawn and the number of the urn from which this ball was drawn. Clearly, there are six different attributes within the original collective, namely, white and No. 1, white and No. 2, white and No. 3, black and No. 1, etc. The corresponding six probabilities are given. Now assume that in a particular case the ball drawn was white, while the number of the urn from which it was drawn is unknown. In that case we may wish to calculate the probability that the ball was drawn from urn No. 1 or, in other words, that the appearance of a white ball was due to the cause that the urn selected was that bearing No. 1. The solution is exactly along the same lines as before: The initial probability of the attribute 'white and No. 1' has to be divided by the sum of the probabilities of the three attributes, white and No. 1, white and No. 2, and white and No. 3. The usual metaphysical formulation of this problem can only be explained historically. The partition rule was first derived by Thomas Bayes,[4] in the middle of the eighteenth century, and his original formulation has since been reprinted in most textbooks practically without alteration.

FORMULATION OF THE RULE OF PARTITION

At this stage I should like to state definitely that in our theory no difference whatsoever exists between the 'probability of causes' (or 'probability of hypotheses') and the more usual 'probability of events'. Our only subject is collectives, i.e., sequences of observational data with various attributes obeying the two laws of the existence of limiting values of relative frequencies and randomness. In every collective possessing more than two distinct attributes, a partition can be carried out. After this, each of the attributes appearing both in the initial and in the derived collectives has two probabilities—the initial one, i.e., that in the original collective, and the final one, i.e., that in the collective derived by partition. There is no place in this problem for any metaphysical formulation.

Before considering the fourth and last of the fundamental operations, I want to summarize briefly the definition of the partition operation, and the solution of the partition problem:

If a collective involves more than two attributes, then, by means of a 'partition', a new collective may be derived from it in the following way.

Consider the set of attributes of the initial collective and choose a certain group of them; pick out from the initial collective all elements whose attributes belong to the chosen group. The selected elements, with their attributes unchanged, will form a new collective.

The distribution within this new collective is obtained by dividing each initial probability of a selected attribute by the sum of the probabilities of all selected attributes.

FOURTH FUNDAMENTAL OPERATION: COMBINATION

The three fundamental operations described so far—selection, mixing, and partition—have one thing in common. In each of them a new collective was derived from *one* original collective by applying a certain procedure to its elements and attributes. The fourth and last operation, which we are now going to consider, is characterized by the fact that a new collective is formed from *two* original ones.

During the discussion of this operation, we shall at the same time gain a first insight into the different forms of relations between two or more collectives. I call this fourth operation *combination*. The example which we are going to use to explain this operation will be, as far as possible, similar to our previous examples of the game of dice. The two initial collectives are now two series of casts; the attributes are in both cases the numbers 1 to 6. The corresponding two sets of six probabilities each, which are not necessarily identical sets, are assumed to be known. The final collective consists of a sequence of simultaneous casts of both dice, and the attributes are the possible combinations of the numbers on both dice. For example, we ask for the probability of the result '3 on the first die and 5 on the second die'. We consider the two dice in this problem to be distinguished by being marked with the figures I and II or having different colours, or other distinctive marks.

Those among you who have learned the elements of the theory of probability at school or have given thought to this problem, know how it can be solved in a primitive way. You will say that it is the question of a probability of 'this as well as that', and the rule is the multiplication of probabilities. If, say, the probability of casting 3 with the first die is 1/7, and that of casting 5 with the second die is 1/6, the probability of casting 3 and 5 with both dice is $1/7 \times 1/6 = 1/42$. This becomes obvious if one thinks that only 1/7 of all casts with the first die are to be taken into consideration, and that in 1/6 of these selected casts the second die is expected to show the number 5.

Clearly, however, this rule requires exact statement and foundation before its general validity can be accepted—a clarification of the same type as was previously given for the addition rule of the probability of 'either-or'. The probability of casting with two dice the sum 8 as well as the difference 2 is, for instance, surely not equal to the product of the two corresponding single probabilities. I shall now consider these finer points, and, in order to be able to present this investigation in a more concise form, I shall use a few simple algebraic symbols. I do not think that this will make these arguments too difficult to follow.

A NEW METHOD OF FORMING PARTIAL SEQUENCES: CORRELATED SAMPLING

We consider a game in which two dice are cast. The method of casting is irrelevant; the dice may be cast from two boxes or from one common box, simultaneously or consecutively. The only essential point is the possibility of establishing a one-to-one correspondence between the casts of die I and those of die II. We consider first only the results obtained with die I. Among the first n of them there will be a certain number, say n_3, of casts in which 3 is the number of points that appeared on the face of the die. The ratio n_3/n is the relative frequency of the result 3 in casting die I; the limiting value of this fraction n_3/n is the probability of casting 3 with this die.

Now we go a step further: Each time that die I has produced the result 3, we note the result of the corresponding cast of die II. This second die will likewise produce the results 1 to 6, in irregular alternation. A certain number of them, say n'_5, will show the result 5.

The relative frequency of the result 5 for the second die (in these selected casts) is thus n'_5/n_3. As we consider now only a partial sequence, derived from the complete sequence of all casts by means of the condition that the corresponding cast of die I should have the result 3, the relative frequency n'_5/n_3 is not necessarily equal to the frequency of 5 in the collective composed of all casts with the second die.

This kind of selection of a partial sequence of the elements of a collective is new to us. The process is different both from place selection, where the elements are selected by means of a pre-established arithmetical rule, independent of their attributes, and from partition, in which the elements selected are those possessing a certain specified attribute. We need therefore a special term to denote this new operation, and we use for this purpose the expression *correlated sampling*, or *sampling* for short. We will say, for instance, that the second collective was sampled by means of the first one, or, more exactly, by the appearance of the attribute 3 in the first collective. In this connexion it will be convenient to use the expressions the *sampled collective* and the *sampling collective*. The procedure may then be described as follows: We start by establishing a one-to-one correspondence between the elements of the collective to be sampled and those of the sampling collective. This is done in our example by casting the two dice each time simultaneously. Next, we choose an attribute of the sampling collective (here the 3 of die I) and select those elements of the sampled collective which correspond to elements of the sampling collective bearing the chosen attribute. In the above example, the first die may be used in 6 different ways to sample the casts with the second die, namely, by means of the attribute 1, attribute 2, . . ., etc.

MUTUALLY INDEPENDENT COLLECTIVES

The ratio n'_5/n_3, which we considered in the preceding paragraph, is the relative frequency of the result 5 in the collective II which is sampled by means of the attribute 3 of the sampling collective I. So far, we do not know the numerical value of this ratio. We are going to assume that by indefinitely increasing the length of the sequence of observations, the ratio n'_5/n_3 tends to a limiting value. But what value? It is possible that this value is equal to the limiting value of the relative frequency of the attribute 5 in the complete sequence of casts carried out with die II. 'Possible', I say, but not certain, since this does not follow from anything we have learned so far. Let us assume for the moment, that it is so. This assumption is simple to understand; it implies that the effect of sampling a partial sequence out of the elements of collective II is similar to that of a place selection, causing no change in probabilities at all. Is there, in fact, any ground for suspecting that the chance of casting 5 with die II may be influenced by the fact that we reject all casts in which die I did not show 3 points? A sceptic may answer: 'Perhaps! It all depends on whether or not the casts of die II are independent of those of die I.' But what does the word 'independent' mean in this connexion?

We can easily indicate conditions under which two collectives are certainly not independent. Two dice tied together by a short thread cannot be expected to produce two sequences of independent results. However, to obtain a *definition* of independence, we must return to the method that we have already used in defining the concepts 'collective' and 'probability'. It consists in choosing that property of the phenomenon which promises to be the most useful one for the development of the theory, and postulating this property as the fundamental characteristic of the concept which we are going to define. Accordingly, subject to a slight addition to be made later on, we now give the following definition: A collective II will be said to be independent of another collective I if the process of sampling from collective II by means of collective I, using any of its attributes, does not change the probabilities of the attributes of collective II, or, in other words, if the distribution within any of the sampled collectives remains the same as that in the original collective II.

If we now assume that the two dice in the above example are independent in the meaning of the word just given, then our problem of finding the probability of the combined attribute (3, 5) can be solved readily.

DERIVATION OF THE MULTIPLICATION RULE

We have considered altogether n casts of two dice and we have found that in n_3 of them the first die showed the attribute 3. Again, among those n_3 casts there were n'_5 such casts in which the second die had the attribute 5. Hence the total number of casts bearing the combined attribute 3 and 5 was n'_5. The relative frequency of this attribute is therefore n'_5/n and the limit of this ratio is just the probability we are looking for. Everybody familiar with the use of mathematical symbols will understand the equation:

$$\frac{n'_5}{n} = \frac{n'_5}{n_3} \times \frac{n_3}{n}.$$

In other words, the relative frequency n'_5/n is the product of the two relative frequencies n'_5/n_3 and n_3/n, both of which we have considered previously. The limiting value of the second of them is the probability of a cast 3 with the first die; we denote it by p_3. According to our assumption of independence of the two dice, the ratio n'_5/n_3 has the same limiting value as the relative frequency of 5 in the complete sequence of casts of die II; in other words, its limiting value is the probability of casting 5 with the second die. Let us denote the probabilities corresponding to die II by the letter q, e.g., the probability of casting 5 by q_5. According to a mathematical rule, the limiting value of a product is equal to the product of the limiting values of the two factors; the limiting value of n'_5/n is thus the product $p_3 \times q_5$. In words: the probability of casting simultaneously 3 with the first die and 5 with the second die is the product of the probabilities of the two separate events. Using the letter P to denote the probabilities in the game with the two dice, we can write the following formula:

$$P_{3,5} = p_3 \times q_5.$$

Analogous formulæ will be valid for all other combinations of two numbers, from 1,1 to 6,6. For instance, the probability of casting 5 with die I and 3 with die II is $P_{5,3} = p_5 \times q_3$, where p_5 denotes the probability of casting 5 with die I in collective I, and so on.

We have introduced and used the definition of independence of collective II with respect to collective I by postulating that the process of sampling from II by means of I should not change the original probabilities in II. However, we have to add that the 'independence' thus defined is actually a reciprocal property; in other words, if II is independent of I, then I is also independent of II. They are mutually independent. This follows from our last formula for $P_{3,5}$, where the two probabilities p_3 and q_5 play exactly the same role. The same argument can be repeated, starting now with collective II and sampling from I by means of the attribute 5 of II, etc. Whenever in what follows we speak of the independence of collective II with respect to collective I, it is with the understanding that the roles of I and II might also be interchanged.

To state the multiplication rule of probabilities for independent collectives in accordance with the general principles of our theory, one more addition must be made. We must be sure that the new sequence of elements formed by the game of two dice, with two numbers as a combined attribute, is a collective in the sense of our definition. Otherwise, no clear meaning would be conveyed by speaking of the probability of the result 3,5. The first criterion—the existence of limiting values—is obviously satisfied, since we have been able to show how these limiting values ($P_{3,5}$ or any other of the thirty-six values from $P_{1,1}$ to $P_{6,6}$) can be calculated. We must now investigate the question of the insensitivity of the limiting values to place selection. To be able to prove the insensitivity we are in fact obliged to add a certain restriction to our previous definition of independence. We must require expressly that the values of the limiting frequencies in collective II shall remain unchanged when we first make an arbitrary place selection in collective I and then use this selected partial sequence of collective I for sampling collective II. The actual need to impose this condition will be illustrated later by an example.

To conclude this section we give a summary concerning the combination of independent collectives.

1. We say that collective II is independent of collective I if the distribution in II remains unchanged by the operation which consists of first, an arbitrary place selection in I, then a sampling of II by means of some attribute in the selected part of I, and finally an arbitrary place selection in the sampled part of II.

2. From two independent collectives of this kind, a new collective can be formed by the process of 'combination', i.e., by considering simultaneously both the elements and the attributes of the two initial collectives.

The result of this operation is that the distribution in the new collective is obtained by multiplying the probabilities of the single attributes in the two initial collectives.

TEST OF INDEPENDENCE

We have thus defined the fourth and last method of forming new collectives. We have merely to add a few words concerning the combination of nonindependent collectives. Before doing this, I will insert another short remark.

In the same sceptical spirit in which we have discussed the concept of probability, we may now ask: How do we know that truly independent collectives exist, i.e., those where the multiplication rule applies? The answer is that we take our conviction from the same source as previously, namely, from experience. As always in the exact sciences, when conclusions are drawn from abstract and idealized assumptions, the test of the value of these idealizations is the confirmation of these conclusions by experiment. The definition of an elastic body in mechanics states that at all points of such a body, strain and stress determine each other uniquely. If we assume that such bodies exist, mechanics teaches us, for instance, how to calculate the deformation by a given load of a girder made of elastic material. How do we know that a particular girder is in fact elastic (in the sense of the above definition), and that therefore the results of the theoretical calculation apply to it? Is it possible to measure the rates of strain and stress at each point of the girder? Obviously not. What we have to do is to assume that the definition applies, calculate the deformation of the girder according to the theory, and test our result by means of an experiment. If satisfactory agreement between theory and experiment is obtained, we consider the premises of the calculation to be correct, not only for the one girder tested, but for all girders (or other objects) made of the same material.

Another and still simpler example is this: Geometry teaches different propositions concerning the properties of a sphere. How do we know that these propositions apply to the earth? Has anybody ever confirmed by direct measurement the existence within the earth of a point equidistant from all points on its surface (this being the geometric criterion of spherical shape)? Surely not. To assume the spherical shape of the earth was first an intuition. This assumption was afterwards confirmed by checking a great number of conclusions drawn from it against empirical results. Finally, slight discrepancies between the theoretical predictions and the experimental results were detected and showed that the sphere is only a first approximation to the true shape of the earth.

Exactly the same conditions are encountered in the case of independent collectives. If two dice are connected by a short thread, nobody will assume mutual independence of the two corresponding collectives. If the thread is somewhat longer, the answer is less obvious, and the best thing to do is to postpone judgment until a sufficiently long sequence of trials has been carried out and the multiplication rule has been tested in this way. If the dice are put into the box singly, without anything connecting them, long-standing and wide experience has demonstrated the validity of the multiplication rule in cases of this kind. If, finally, the two dice are thrown by two different persons from separate boxes, perhaps even in two distant places, the assumption of independence becomes an intuitive certainty, which is an outcome of a still more general human experience. In each concrete case, however, the correctness of the assumption of independence can be confirmed only by a test, namely, by carrying out a sufficiently long sequence of observations of the dice under consideration, or of another system considered to be equivalent to the one in which we are interested. The results of this test are compared with the predictions deduced from the assumption of the multiplication rule, and the collectives are considered as independent if agreement between theory and experiment is found.

The mutual independence of two collectives can often be deduced directly from their definition. This is true when both collectives are derived in a certain way from a common original collective. Examples of this kind will be discussed later when we deal with the repeated use of the four fundamental operations.

COMBINATION OF DEPENDENT COLLECTIVES

To conclude this discussion, I shall briefly describe how the combination of two collectives operates in cases, in which the conditions of independence are not completely satisfied. We do not mean cases in which no condition whatsoever is imposed on the two collectives; far from it. It is only a matter of a slight relaxation in the condition of independence. We shall say that two collectives *A* and *B* are *combinable* but *interdependent* if the following relation exists: We start, as before, with an arbitrary place selection in *A*. Next we use, as before, some attribute within this selected sequence in order to sample a partial sequence of *B*. In contrast to the previous definition of independence, we assume now that the distribution of probabilities in the sampled partial sequence of *B* depends on the attribute in *A* that was used for the sampling. Here is a concrete example: The dice *A* and *B* are cast simultaneously. The probability of obtaining 5 with *B*, if we count only those casts where *A* has given the result 3, has a definite value. This value is, however, now assumed to be different from the probability of obtaining 5 with *B* if the sampling is made by means of the result 4 for *A*. The following is an illustration.

Three black balls and three white balls are placed in an urn. We draw two balls consecutively, without putting the first ball back before drawing the second one. The two balls are then replaced and the whole procedure is repeated. The first of the two collectives under consideration is that composed of all the 'first' draws, i.e., draws made from the urn containing six balls; the probability of a white ball in this collective is 1/2, if the distribution in the urn is uniform. The second collective consists of all 'second' draws, out of the urn containing only five balls. This second collective can be sampled by means of the first one. Two partial sequences of elements are obtained in this way; the first contains all second draws following the

55

drawing of a white ball, and the second contains all second draws following the drawing of a black ball. The probability of drawing black is 3/5 in the first of these two new collectives, and only 2/5 in the second one. This follows from the fact that in the first case all three black balls remained in the urn after the first ball had been drawn, whereas in the second case, the number of black balls left after the first draw was only two. The distribution of probabilities in the sampled collectives depends in this case on the attribute of the first collective used for the sampling. It can easily be seen how the final distribution can be calculated in such a case of combinable, but not independent, collectives. To obtain, e.g., the probability of the sequence black ball–white ball, one must multiply the following two factors: the probability 1/2 of a first black ball and the probability of a second ball being white calculated under the assumption that the first one was black. This last probability is 3/5; the result is therefore 1/2 × 3/5 = 3/10. Analogous calculations can be carried out for all other combinations of the two properties black and white.

EXAMPLE OF NONCOMBINABLE COLLECTIVES

It is, finally, not without interest to give an example of two collectives which are neither independent nor dependent in the above sense, collectives which we consider altogether uncombinable. Imagine that a certain meteorological quantity, such as the relative humidity of the air, has been measured over a long time every day at 8 a.m. The results are expressed by numbers, say by integers from 1 to 6. These numbers are the attributes in the collective formed by these consecutive measurements. Now imagine that the same or another meteorological quantity has also been measured every day at 8 p.m. This gives us a second collective, the elements of which are in one-to-one correspondence with the elements of the first collective. We assume that both sets of measurements have the essential properties of collectives, namely, existence of limiting frequencies and randomness. We can, furthermore, assume that the distribution in the second collective is not affected by sampling by means of the first one, in other words, that evening measurements following a certain morning value, say the value 3, have the same distribution as those following any other morning result. All these assumptions do not preclude, however, the possibility of a regularity of the following kind: on each 28th day a morning value, 3, if it happens to occur on this day, automatically involves the occurrence of the same value 3 in the evening. In a case like this, the combination of the two collectives produces a sequence which is not a collective. By applying to the first collective the place selection consisting only of the 28th, 56th, 84th . . . observations, we obtain a sequence for which the probabilities of the attribute combinations (3,1), (3,2), (3,4), (3,5), and (3,6) are zero. For example, $p_{3.1} = p_3 \times 0 = 0$, where p_3 is the probability of the morning value 3, and 0 the probability of an evening value 1 following a morning value 3. The probability of the combination (3,3), i.e., $p_{3.3}$ equals $p_3 \times 1 = p_3$, since the probability of an evening value 3 following a morning value 3 is 1. The distribution in the selected sequence is thus different from that for the total of all morning and evening data, which shows for all possible combinations definite nonzero probabilities. The sequence of elements obtained by combination is, in this case, not a collective, since in this sequence the limiting values of the relative frequencies can be changed by place selection. The initial single sequences of observations have the property of randomness; they have, however, a certain mutual relation which precludes their being combined into a new collective. We call two collectives of this kind *noncombinable*.

This example illustrates again the insufficiency of the well-known elementary form of the multiplication rule, which does not take into account the possible relations between the two collectives. A reliable statement of the multiplication rule can only be based on a rational concept of probability whose foundation is the analysis of the collective.

I shall give the substance of what we have learned about the four fundamental operations, in the form of the following short statements:

1. *Selection:* Definition: The attributes unchanged, the sequence of elements reduced by place selection. Solution: The distribution is unchanged.

2. *Mixing:* Definition: Elements unchanged, attributes 'mixed'. Solution: Addition rule.

3. *Partition:* Definition: Attributes unchanged, sequence of elements reduced by partition. Solution: Division rule.

4. *Combination:* Definition: Attributes and elements of two collectives combined in pairs. Solution: Multiplication rule.

With the statement of these four fundamental operations, and with the indication of the methods for determining the distributions in the derived collective from that (or those) in the initial ones, the foundations of the theory of probability are laid. The method of solving concrete problems by the application of this general scheme is as follows:

First of all, we must find out what the initial collectives are and state their distributions. Secondly, we must consider the final collective whose probabilities we are asked to determine. Finally, we have to carry out the transformations from the initial collectives to the final one, in steps which consist of the fundamental operations. The problem is then solved by applying to each operation its solution from the above scheme. Of course, it is not always necessary to proceed pedantically in this way, especially after one has acquired a certain experience. An experienced worker in the field immediately recognizes certain connexions between the collectives under consideration. He will use certain repeatedly occurring groups of fundamental operations as new operations, which he applies in one step. In many examples, not only in the simplest, the entire preparatory work reduces to a minimum. The whole solution may, for instance, consist in a single mixing operation; this may, however, involve difficulties of a purely analytic nature, consisting in the evaluation of complicated sums or integrals.

In the following sections I shall discuss an example in which no mathematical difficulties are involved, but which illustrates several combinations of the fundamental operations.

A PROBLEM OF CHEVALIER DE MÉRÉ

This is perhaps the oldest problem ever solved by probability calculus; a consideration of it will be useful for us from more than one point of view.

In the time of Pascal and Fermat,[5] two great seventeenth-century mathematicians, there lived in France a certain Chevalier de Méré, a passionate gambler. One of the games of chance fashionable in his time was played in this way: A die was cast four times in succession; one of the players bet that the 6 would appear at least once in four casts; the other bet against it. Chevalier de Méré found out that there was a slightly greater chance of getting the positive result (i.e., 6 coming out at least once in four casts). Gamblers sometimes like variety and the following variation of the game was introduced: Two dice were used instead of one, and were thrown twenty-four times; the subject of the betting was the appearance or nonappearance of at least one double 6 in twenty-four casts. Chevalier de Méré, who was obviously a studious gambler, found out that in this case the win went more often to the player betting against the positive result (the appearance of the combination 6,6). This seemed to him strange and he even suggested that arithmetic must be wrong in this case. His argument went as follows: The casting of a single die can produce six different results, that of two dice thirty-six results, six times as many. One of the six possibilities in the game with one die is 6; one of the thirty-six possibilities in the game with two dice is the combination 6,6. In casting two dice twenty-four times, the chance of casting 6,6 must be the same as that of casting 6 in four casts of one die. Chevalier de Méré asked Fermat for a solution of this paradox; and Fermat solved it. His solution has been preserved for us in a letter addressed to Pascal.

I will give the solution in the following section, in a more general form than that given by Fermat, and will show how the solution follows from the concepts on which we have founded the theory of probability.

SOLUTION OF THE PROBLEM OF CHEVALIER DE MÉRÉ

We begin with the simpler case, that of the four casts with one die. The initial collective is obviously the sequence of casts with one die; the elements are single casts, and the attributes are the numbers 1 to 6. Fermat assumed implicitly that the six attributes are equally probable, i.e., that the die used is an 'unbiased' one; this assumption, which assigns the value 1/6 to each of the six probabilities, forms the basis of his calculations. According to our general concepts, the solution can be found without this special assumption. We ascribe to the six possible results the six probabilities $p_1, p_2, \ldots, p_6$, which may be all equal or different from each other, but in any case give the sum 1.

What is the problem? We are asked to determine the probability of a 6 appearing at least once in a group of four casts. This is obviously a probability in the following new collective: the elements are groups of four consecutive casts; the attributes are 'yes' or 'no' (simple alternative)—'yes', if at least one of the four results is 6, 'no' if no 6 occurs in these four results. This is the collective which we must derive from the initial one. What we are interested in is the probability of the attribute 'yes' in this final collective.

We must now find out which of our fundamental operations lead to the final collective K from the initial one, which we may denote by C. First of all, we drop the distinction between the results 1, 2, 3, 4, and 5, because we are only asked whether the result is 6 or not.

We begin therefore by mixing the attributes 1 to 5, and leaving 6 as a second alternative attribute. We form in this way a new collective, which we may call C', consisting of the same elements as C, but with only two attributes, 6 and not-6. According to the addition rule, the probabilities of these two attributes are

$$p_6 \text{ and } p_1 + p_2 + p_3 + p_4 + p_5,$$

respectively. Since the sum of these two probabilities must be 1, we can replace the last sum by $(1 - p_6)$.

We now apply to C' a selection, by selecting from the infinite sequence those elements whose numbers in the original sequence are

$$1, \ 5, \ 9, \ 13, \ 17, \ 21, \ 25 \ . \ . \ .$$

The attributes in this new collective—let us call it C'_1—are the same as in C' (i.e., 6 and not-6). According to our general rules, the distribution must also be the same, and the probability of a 6 is therefore still p_6, that of not-6 is $(1 - p_6)$.

We can form a second similar collective by another selection from C', namely, by retaining the elements whose numbers are

$$2, \ 6, \ 10, \ 14, \ 18, \ 22, \ 26 \ . \ . \ .$$

We call this collective C'_2; again, the probability of 6 in it is p_6 and that of not-6 is $(1 - p_6)$.

In the same way we carry out a third selection, that of the elements

$$3, \ 7, \ 11, \ 15, \ 19, \ 23, \ 27 \ . \ . \ .,$$

and a fourth selection—that of the elements

$$4, \ 8, \ 12, \ 16, \ 20, \ 24, \ 28 \ . \ . \ .$$

These last two collectives we call C'_3 and C'_4. We have thus formed altogether four new collectives, C'_1, C'_2, C'_3, and C'_4 by selection from the collective C'; the attributes in each of them are simple alternatives with the probabilities p_6 for the attribute 6, and $(1 - p_6)$ for the attribute not-6. These probabilities are known quantities, since we assumed that the values of $p_1, p_2, \ . \ . \ ., p_6$ are the given data of the problem.

It remains now to make one last step: to carry out a combination of the four collectives derived by selection. Let us first combine C'_1 with C'_2; this means that we couple together the first elements of the two collectives (casts 1 and 2), the second ones (casts 5 and 6), the third ones (casts 9 and 10), and so on. The new collective formed in this way we call C''_1; its elements are certain pairs of casts, namely, those pairs whose places in the original sequence were

1 and 2; 5 and 6; 9 and 10; 13 and 14; 17 and 18; . . .

The attributes in this collective are the four possible combinations of the two attributes 6 and not-6—i.e., 6 and 6, 6 and not-6, not-6 and 6, and not-6 and not-6.

Are we right in applying the procedure of the combination of independent collectives to the collectives C'_1 and C'_2? The answer is in the affirmative; this case is one of those mentioned above, in which the independence follows directly from the derivation of the collectives. This fact can be proved mathematically; it is, however, easy to recognize, without any mathematical deduction, that the randomness of the initial collective C implies the mutual independence of the collectives C'_1 and C'_2 derived from it (via the intermediate collective C'). The probabilities of the four combinations mentioned above can therefore be calculated by means of the multiplication rule. That of the first one (6 and 6) is p_6^2, that of the second and third one (6 and not-6, not-6 and 6) is $p_6(1 - p_6)$, and that of the fourth one (not-6 and not-6) is $(1 - p_6)^2$.

Exactly the same kind of combination can be carried out with the collectives C'_3 and C'_4. The new collective, C''_2, formed in this way, contains the following pairs of casts:

3 and 4; 7 and 8; 11 and 12; 15 and 16; 19 and 20; . . .

The attributes and the probabilities are the same as in C''_1.

We now proceed to the last *combination*—that of C''_1 and C''_2. This process means the coupling together of two pairs, e.g., casts 1 and 2 (from the collective C''_1) with casts 3 and 4 (from the collective C''_2), next casts 5 and 6 with casts 7 and 8, and so on. The elements of the new collective are thus groups of four casts each, those numbered

1 to 4; 5 to 8; 9 to 12; 13 to 16; 17 to 20; . . .

We denote this collective by K'; its attributes are the sixteen possible combinations of the four attributes occurring in C''_1 with the four attributes occurring in C''_2. The corresponding sixteen probabilities can be derived by the multiplication rule, whose applicability in this case is due to the same relations as in the case of the combination of C'_1 with C'_2, and of C'_3 with C'_4. The probability of the attribute '6 and 6', '6 and 6' (four '6's), for instance, is $p_6^2 \times p_6^2 = p_6^4$; that of the attribute 'four times not-6' is $(1 - p_6)^4$; and so on.

We are now at the last stage of the calculation, leading from K' to the final collective K. We are not interested in the probabilities of all the sixteen attributes occurring in K', but only in the alternative: no-6 at all, i.e., four times not-6, or all the other results. Another mixing is thus necessary. The probability of the property 'no-6 at all' remains unaffected by mixing, i.e., equal to $(1 - p_6)^4$. The probabilities of the remaining fifteen results need not be calculated separately. Their sum is given by the expression

$$p = 1 - (1 - p_6)^4.$$

This is the probability of the property 'not four times not-6'— i.e., 'at least one 6'—in the collective K (derived by mixing from the collective K'). Our problem is thus solved.

DISCUSSION OF THE SOLUTION

The result can be extended, without much further calculation, to the second part of de Méré's problem—the case of twenty-four casts of two dice. We consider the sequence of casts with two dice as the initial collective C; the result in which we are interested is a double 6. The probability p_6 in the previous derivation must be replaced now by the probability $p_{6,6}$, that of casting a double 6 in an indefinitely long sequence of casts of two dice. The solution is found along exactly the same lines as above, although twenty-four selections are now to be made instead of the four selections required in the simpler example, and twenty-four selected collectives must be combined in successive steps. We need not discuss all these steps in detail; the outcome is simply the substitution of the exponent twenty-four for the exponent 4 in the above-given formula. Hence,

$$p' = 1 - (1 - p_{6,6})^{24}$$

is the probability of a double 6 appearing at least once in a series of twenty-four casts with two dice.

Assuming that the results of the game with two dice can be considered as a combination of two independent collectives, we can express the initial collective C of the second part of the problem in terms of the initial collective C of the first part. The probability $p_{6,6}$ is in this case equal to $p_6{}^2$. The formula for the probability of a double 6 becomes

$$p' = 1 - (1 - p_6{}^2)^{24}.$$

This last probability we now wish to compare with the probability p calculated for the game with one die.

We see, first of all, that the two expressions are different. In other words the values of p and p' are not identical for arbitrary values of the probability p_6. De Méré had surely a correct die in mind, with $p_6 = 1/6$. By introducing this particular value into the two formulæ we obtain the following numerical values of p and p':

$$p = 1 - (5/6)^4 = 0.516,$$
$$p' = 1 - (35/36)^{24} = 0.491.$$

The observations of de Méré were thus correct: in betting on a 6 in four single casts, the chance is somewhat higher than 0.5, and in betting on a double 6 in twenty-four double casts, somewhat lower than 0.5. It is therefore profitable to bet on 'yes' in the first game, and on 'no' in the second one. His reasoning was, however, inexact, and his conclusion that, theoretically, the chances must be the same in the two cases, was wrong.

SOME FINAL CONCLUSIONS

A number of useful consequences may be drawn from this solution. First of all we see that the solution of a problem in the theory of probability can teach us something definite about the real world. It gives a prediction of the result of a long sequence of physical events; this prediction can be tested by observation. Historically, in this case the observation preceded the calculation. This is, however, of no basic importance, since the result of the calculation is of general validity and can be applied to all similar cases in the future. For instance, another number may be substituted for 4 or 24, a biased die can be used instead of the unbiased one (i.e., the probability p_6 can be different from 1/6), etc. Another characteristic property of results obtained by probability calculus clearly illustrated by this problem is that all such results apply to relative frequencies of events in long sequences of observations, and to nothing else. Thus, a probability theory which does not introduce from the very beginning a connexion between probability and relative frequency is not able to contribute anything to the study of reality.

I should like to stress here also another side of the problem under discussion. It is often assumed that in games of chance we are always dealing with probabilities known a priori from the general principle of equal probability of all possible results. However, in the game considered in the above example, there is no reason for assuming a priori that the chances of the two partners are the same. It is by no means obvious that exactly four casts of a die should be necessary to give to the result 6 the chance 0.5. The origin of such games can only have been the observation of a very large number of actual casts of dice.

The history of this particular game of dice might have been as follows: In the course of centuries, men learned how to make dice so that the chances of all six results were about the same. Afterwards, it was found that with these unbiased dice the chance of casting 6 once in four casts was close to 50%, and that the same was true for the chance of casting a double 6 in twenty-four casts. Finally, longer series of observations showed that in these two cases the probabilities were not exactly equal to 0.5; deviations were found which required explanation. Now came the theory, which investigated the relation between the two properties of a die—namely, its property of falling equally often on each of its sides, and its property of giving in half of the sequences of four casts at least one 6. Calculations showed that these two properties are not strictly in accordance with each other: The value $p_6 = 1/6$ results in a value $p = 0.516$ and not 0.5. It is also easy to calculate, from the above formula, that $p = 0.5$ requires, conversely, a value of p_6 slightly smaller than $1/6$, namely 0.1591. It is hardly possible to demonstrate more clearly the empirical character of the theory of probability and its purpose of interpreting observable phenomena.

However, this has brought us to the subject-matter of the next lecture. I shall not, therefore, pursue this line any further here. I must also abstain from considering more examples and from discussing more special problems in detail. They would teach us little that we do not already know.

On a former occasion I said that experience in calculating helps us to simplify the solution of special problems, to reduce the number of necessary steps, which was so large in the example discussed in the preceding paragraphs. It is, however, not my task to give you this practical tuition. I prefer to close by giving a short summary of the most important points, marking our development of probability calculus.

SHORT REVIEW

1. The starting point of the theory is the concept of a collective. Probability has a real meaning only as probability in a given collective.

2. A collective is an infinite sequence of observations, each

observation ending with the recording of a certain attribute. The relative frequency with which a specified attribute occurs in the sequence of observations has a limiting value, which remains unchanged if a partial sequence is formed from the original one by an arbitrary place selection.

3. The limiting value of the relative frequency of a given attribute —which, as just stated, is insensitive to all kinds of place selections— is called its probability within the collective considered. The probabilities of all the attributes within a collective form its distribution.

(This much was covered in the first lecture. Now we come to the new matter we have learned in the second lecture.)

4. The task of the theory of probability is to derive new collectives and their distributions from given distributions in one or more initial collectives. The special case of a uniform distribution of probabilities in the original collective ('equally probable' cases) plays no exceptional role in our theory.

5. The derivation of a new collective from the initial ones consists in the application of one or several of the four fundamental operations (Selection, Mixing, Partition, Combination).

6. The first operation, Selection, leaves the distribution unchanged; the second one, Mixing, changes it according to the addition rule; the third one, Partition, changes it according to the division rule, and the fourth one, Combination, changes it according to the multiplication rule.

7. The knowledge of the effect of the four fundamental operations on the distribution enables us, in principle, to solve all problems of the calculus of probabilities. Actual problems may be, nevertheless, very nvolved, whether on account of difficulties in the logical analysis of the problem, i.e., in the enumeration of the necessary elementary operations; or because of complications arising from the accumulation of a great number of elementary operations; or finally, because of purely analytical difficulties.

8. Each probability calculation is based on the knowledge of certain relative frequencies in long sequences of observations, and its result is always the prediction of another relative frequency, which can be tested by a new sequence of observations.

The following is a summary of the essence of these points in a single sentence:

The theory of probability deals exclusively with frequencies in long sequences of observations; it starts with certain given frequencies and derives new ones by means of calculations carried out according to certain established rules.

Critical Discussion of the Foundations of Probability

I HAVE given, in the first two lectures of this series, an outline of what I call the foundation of the new probability theory. The main points of the theory were briefly restated at the end of the last lecture. If it were my intention to give a complete course on the theory of probability, I should now demonstrate how new collectives are derived from given ones by more and more complicated combinations of the four fundamental operations, and, on the other hand, how all problems usually treated in probability calculus can be reduced to combinations of this kind. It would, however, be impossible to do this without using mathematical methods out of place in this book. Those who are interested in this side of the theory may refer to my *Lectures on the Theory of Probability*, originally published in 1931, and to my *Mathematical Theory of Probability and Statistics* of 1946 (see Notes and Addenda, p. 224). Here, we are interested in the general foundations of the theory.

This lecture will therefore deal with a critical survey of the results described in the first two lectures. Discussion will proceed along two lines. First, I shall consider the relation of the new theory to the classical one[1] and to some of the recent modifications which are intended to provide the classical theory with a firmer foundation. Second, I am going to deal with the numerous works which have appeared since my first publications which have sometimes contained objections to my theory, and sometimes suggestions for its modification or further development.

THE CLASSICAL DEFINITION OF PROBABILITY

The 'classical' definition of probability was given by Laplace and has been repeated, until about 1930, in nearly all the textbooks on the theory of probability with its form almost unchanged. It runs:

66

Probability is the ratio of the number of favourable cases to the total number of equally likely cases. This same idea underlies all the work prior to that of Laplace, although the authors did not always state the definition clearly.

I must point out, however, that in more recent times many mathematicians have been aware of the inadequacy of Laplace's definition. For instance, Poincaré[2] (1912) says: 'It is hardly possible to give any satisfactory definition of probability; the usual one is as follows . . .' Later, we shall see that a complete logical development of the theory on the basis of the classical definition has never been attempted. Authors start with the 'equally likely cases', only to abandon this point of view at a suitable moment and turn to the notion of probability based on the frequency definition; in fact, they even sometimes explicitly introduce a definition of this kind. For this reason I maintain that the gulf between the new point of view and the classical one is not unbridgeable. For most mathematicians the acceptance of the new definition would only mean the surrender of the form in which the theory of probability is usually presented, namely, one which permits the solution of a number of simple problems at the beginning of the course, while avoiding the immediate discussion of more difficult and fundamental ideas.

The main objection to the definition given by Laplace concerns the expression 'equally likely' or 'equally possible cases'. Ordinary speech recognizes different degrees of possibility. A certain event is sometimes called 'possible' or 'impossible'; it may equally well be 'quite possible' or 'hardly possible', and these expressions mean that we are conscious of the varying degrees of 'effort' involved. It is 'hardly possible' to write, in longhand, at a speed of forty words a minute, and 'impossible' at a speed of a hundred and twenty words a minute. Nevertheless, it is 'quite possible' to achieve the first speed with a typewriter, and the second by using shorthand. In the same way we call two events 'equally possible' if the same 'effort' is required to produce each of them. Jacob Bernoulli,[3] a predecessor of Laplace, does in fact speak of events 'quod pari facilitate mihi obtingere possit' (that can be achieved with equal ease). However, Laplace was thinking of something else when he spoke of 'equally likely cases'.

In another sense we say, 'This event is more likely than that', and in this way we express our conjecture concerning what we expect to occur; this is the sense in which Laplace and the followers of the classical theory of probability use the phrase 'equal possibility'. Thus we see that this latter phrase merely means 'equally reliable conjectures' or, to use the current expression, 'equal probabilities'. The phrase 'equally likely cases' is exactly synonymous with 'equally probable cases'. Even such a voluminous treatise as A. Meinong's[4] *Probability and Possibility*, only serves to confirm this simple fact. If we remember that with equal probabilities in a collective the distribution was called *uniform*, we may say that, unless we consider the classical definition of probability to be a vicious circle, this definition means the reduction of all distributions to the simpler case of uniform distributions.

EQUALLY LIKELY CASES . . .

We must now examine a little more closely the way in which this reduction is carried out. An unbiased die can produce six equally likely results. One of them is the number 3, and, therefore, the probability of throwing a 3 is 1/6. If a wheel used in a lottery bears the numbers 1 to 90, there are ninety equally likely cases. Nine of these correspond to numbers expressed by a single digit (1 to 9); nine others are two-digit numbers which are divisible by 10; the remaining seventy-two numbers have two digits and are not divisible by 10. Therefore, the probability of a number with a single digit is 9/90 = 1/10, and so is that of a number divisible by 10; the probability of all the other results is 8/10. Let us consider as a third example a game played with two unbiased dice. Each possible combination of numbers is an equally likely case; there are thirty-six such combinations. Therefore the probability of throwing a double 6 is 1/36; that of throwing the sum 11 is 1/18, because two cases are favourable to this result, namely, 5 on the first die and 6 on the second, and vice versa.

The consideration of these three applications of the theory of 'equally likely cases' leads to the following conclusions. In the first case, we have obviously a mere tautology, if we remember that the expressions 'equally likely' and 'equally probable' are identical, the only other consideration involved being that the sum of all the probabilities is equal to unity. In the second case, we have several favourable cases united into a group. This is a special case of the operation discussed in the previous lecture, that of mixing the labels in a collective. The attributes 1, 2, 3, . . ., 9 are mixed, the attributes 10, 20, . . ., 90 likewise, and also the remaining attributes. Three groups of attributes are thus formed, the single probabilities being each equal to 1/90; the addition of the probabilities in each group produces the results shown above. For the game with two dice which formed our third example, the classical theory uses a theorem relating to the combination of independent collectives. In a very specialized form this theorem states that: Each combination of one of the equally likely cases from the first and from the second collectives produces an equally likely case in the new collective. As we know, we can solve the problems which occur most frequently and are most important in the theory of probability by the processes of mixing and combination. Hence, the theory of equal possibilities permits us to solve most problems in which there are uniform distributions of probabilities in the original collectives. Most of the usual games of chance—unbiased dice, properly made roulette wheels, and so forth—produce collectives of this kind.

. . . DO NOT ALWAYS EXIST

But how are we to deal with the problem of a biased die by means of a theory which knows only probability based on a number of equally likely results? It is obvious that a slight filing away of one corner of an unbiased die will destroy the equal distribution of chances. Are we to say that now there is no longer a probability of throwing a 3 with such a die, or that the probability of throwing an even number is no longer the sum of the probabilities of throwing a 2, 4, or 6? According to the classical theory, none of the theorems derived on the basis of equally likely cases can be applied to a biased die (since there is no probability without equally likely cases). Nevertheless Laplace[5] in his fundamental treatise attempted to deal with the case of a coin which had different chances for showing heads or tails. It was later realized that his conclusions were not valid and later textbooks on the theory of probability merely omitted any consideration of these questions. The biased die was not considered a subject worthy of treatment by the calculus of probability. It is obvious that such a point of view admits of no argument.

There are other problems, however, belonging to the same category as the biased die which cannot be set aside so easily. One of these is the problem of the probability of death. According to a certain insurance table (see note 14, lect. 1), the probability that a man forty years old will die within the next year is 0.011. Where are the 'equally likely cases' in this example? Which are the 'favourable' ones? Are there 1000 different possibilities, eleven of which are 'favourable' to the occurrence of death, or are there 3000 possibilities and thirty-three 'favourable' ones? It would be useless to search the textbooks for an answer, for no discussion on how to define equally likely cases in questions of this kind is given. When the authors have arrived at the stage where something must be said about the probability of death, they have forgotten that all their laws and theorems are based on a definition of probability founded only on equally likely cases. The authors pass, as if it were a matter of no importance, from the consideration of a priori probabilities to the discussion of cases where the probability is not known a priori, but has to be found a posteriori by determining the frequency of the different attributes in a sufficiently long series of experiments. With extraordinary intrepidity all the theorems proved for probabilities of the first kind are assumed to be valid for those of the second kind. If an author wishes to substantiate this step, he usually refers to Bernoulli's so-called Law of Large Numbers, which is supposed to form a bridge between the concept of a priori probabilities and the determination of probabilities from observations.

We shall see later that this does not work, and that the whole chain of argument is completely circular. Without awaiting this discussion, we may say at once that, up to the present time, no one has succeeded in developing a complete theory of probability without, sooner or later, introducing probability by means of the relative frequencies in long sequences. There is, then, little reason to adhere to a definition which is too narrow for the inclusion of a number of important applications and which must be given a forced interpretation in order to be capable of dealing with many questions of which the theory of probability has to take cognizance. The position may be illustrated by an analogy from the field of elementary plane geometry.

A GEOMETRICAL ANALOGY

Somebody might consider the possibility of developing the geometry of closed rectilinear figures (polygons) from the exclusive consideration of polygons with equal sides of one given length. In this kind of geometry there would be no measurement of length, all figures being determined by their angles and number of sides. If an adherent of this system of geometry were presented with a triangle having sides of different lengths, say three, four, and five units, he would describe this figure as an equilateral dodecagon in which three, four, and five sides respectively fall on straight lines, i.e., a dodecagon with nine of its angles each equal to 180 degrees. The reduction of all polygons to equilateral ones is possible without great difficulty provided all the sides are multiples of a certain unit length;

this can be carried to any desired degree of accuracy, if a sufficiently small unit is taken. Nevertheless, in a geometry of this kind a distinction must be drawn between polygons for which the number of sides is known *a priori* (all of their angles being different from 180°), and those for which it must be determined *a posteriori* by expressing the lengths of their sides, exactly or approximately, as multiples of the unit length.

It is quite possible to develop a theory of this kind, but no mathematician will say that the concept of length and the measurement of length can be entirely removed from geometry in this way. In fact, such a theory is merely a roundabout way of replacing the more direct approach.

The same holds true for the theory of probability based on equally likely cases. From an historical point of view, it is easy to understand why the theory started with the consideration of cases of equal probability (corresponding to the equilateral polygons). The first subjects of the theory of probability were games of chance based on uniform distributions of probabilities. If, however, a modern mathematician attempts to reduce the probabilities of life and death, determined as relative frequencies, to some hypothetical equally likely cases, he is merely playing hide and seek with the necessity for a comprehensive definition of probability which for our theory is just as unavoidable as the idea of length and of its measurement are for geometry.

HOW TO RECOGNIZE EQUALLY LIKELY CASES

I think that I have made clear the distinction between our definition of probability and the classical one, which is still preferred by a few authors. I anticipate that, in the future, the more important problems of insurance, statistics, and the theory of errors will take precedence over the problems of gambling, which are chiefly of historical importance. Then there will be no hesitation in founding the theory of probability on principles which are both simple and rational. In fact, we have already entered upon this development.

Various authors have asked how it is possible to be sure that each of the six sides of a die is equally likely to appear or that each of ninety numbers in a lottery is equally likely to be drawn. Our answer is of course that we do not actually know this unless the dice or the lottery drums have been the subject of sufficiently long series of experiments to demonstrate this fact. In contrast to this point of view, the defenders of the classical theory use a particular argument to support their concept. They assert that the presence of equally likely cases is a piece of a priori knowledge.

Let us assume that a perfect geometrical cube has been made from *perfectly homogeneous* material. One would think that it is then clear, a priori, that none of the six sides can be more likely to show up than any other. One usually states that in this case it is certain that the chance of the cube falling on any particular side is the same for all six sides. I will concede this statement for the moment, although the result of the throw depends also on the dice box, as well as on the whole process of throwing the die from the box, and so on. I will also forget that the statement has a definite meaning only if we already know what 'equal chance' means. For example, we may have adopted the frequency definition, and according to this, 'equal chance' of a number of results would mean equal frequency in a long series of throws. Without some such definition, the statement conveys no knowledge at all, either a priori or of any other kind. Let us, however, overlook these points and assume that the whole task consists in ascribing some fractions, the 'probabilities', to the six sides of the die. The question arises whether, for an actual cube, we can arrive at the conclusion that all these numbers must be equal by a logical process of thought, independent of experience. As soon as we consider more closely the assumptions of homogeneity and symmetry which must be satisfied by the cube we find out the practical emptiness of the whole statement.

We can call a material 'homogeneous' in a logical sense if no particular distinction can be made between any of its parts; that is, the material must also be one whose parts have the same origin and history. However, one part of the ivory of which the die is made was certainly nearer to the tip of the tusk than some other part; consequently, the identity of behaviour of all parts is no longer a logical necessity. This identity of behaviour follows in fact from experience, which shows that the original position of the ivory on the animal does not influence its properties in this respect.

In a concrete example, we not only use this but many other deductions from experience as well. For instance, we inscribe the six sides of the die with six different numbers and assume that this does not affect the relative chances. Primitive tribes, i.e., human beings with a restricted experience, frequently believe the fate of the human body to be affected by inscriptions on its different parts. Moreover, we not only paint the numbers on the die, but make from one to six incisions on its surface and so substantially change its geometrical symmetry; we still assume, on the basis of experience, that this does not affect the chances of the game. If a supporter of the a priori concept of probability is pressed to explain what he understands by 'complete homogeneity', he finally merely requires that the centre of gravity of the cube should coincide with its geometrical centre. If he knows enough mechanics, he adds that the twelve moments of inertia about its twelve edges must all be equal. No one will any longer maintain that it is evident a priori that just these conditions are necessary and sufficient for the 'equal possibility' of the six sides of the die, and that no further conditions, such as conditions involving moments of higher order, need be considered. In fact, this formulation contains a number of results taken from the mechanics of rigid bodies, a science likewise based on experience. We may sum up our discussion by saying that no concrete case can be handled merely by means of an a priori knowledge of equally likely cases. It is always necessary to use more or less general results derived from observation and experience in order to determine which properties of the apparatus that we are using may influence the course of the experiments, and which properties are irrelevant from this point of view.

The situation is very much the same as in the well-known application of the principle of symmetry to the derivation of the equilibrium conditions for a lever with equal arms. When the two sides of the lever are completely identical, the equality of the forces is assumed to follow 'by reason of symmetry'. This form of the lever theorem is, however, much too specialized; apart from the practical impossibility of constructing a lever with exactly identical sides (in the logical sense we have discussed), we must bear in mind that a lever with equal arms is not defined as one having identical arms, but as one in which the forces act at equal distances from the fulcrum. No further geometrical symmetry is required. It is instructive to see, in the older textbooks of applied mechanics, how many figures representing levers of different shapes have been drawn to acquaint the student with the idea of equal-arm levers which do not possess geometrical symmetry. Yet this decisive fact, that only the distances from the fulcrum matter, is a result of experience and observation.

ARE EQUALLY LIKELY CASES OF EXCEPTIONAL SIGNIFICANCE?

Those who admit the insufficiency of the above-explained a priori approach but wish to maintain the exceptional role of 'equally likely' cases may reason as follows: If, in addition to geometrical symmetry, a cube possesses 'kinetic symmetry' (equal moments of first and

73

second order), then the equal chances for the six faces follow from the mechanics of rigid bodies.

However, let us consider now the case of a biased die; we find that here mechanics gives us no assistance. When we have determined all the mechanical characteristics of this die, centre of gravity, moments of inertia, etc., we are still unable to derive, by means of mechanics, the relative frequencies with which it will fall on its different sides. In this case, the only way to determine the probability of the different results is by statistical experiment. The case of a symmetrical die is thus distinguished from that of an unsymmetrical one in that in the former case a prediction of probabilities is possible, if not a priori, at least by the application of an experimental science (mechanics) which is of a distinctly deterministic character.

I think, however, that there is a flaw in this argument. I have already pointed out that the result of a statistical experiment with a die depends not only on the die but on the whole process of throwing it. It is possible to cheat, wittingly or unwittingly, with a perfectly symmetrical die by using certain tricks in placing the die in the box or throwing it out. Very delicately balanced psychological or physiological phenomena are sometimes involved in these procedures. This is well known from the experience with card sharps as well as from certain observations which have often defied explanation and are the favourite subject-matter of so-called 'parapsychology'.[6] I do not want to defend the occult sciences; I am, however, convinced that further unbiased investigation of these phenomena by collection and evaluation of old and new evidence, in the usual scientific manner, will lead us sooner or later to the discovery of new and important relations of which we have as yet no knowledge, but which are natural phenomena in the usual sense. At any rate, it is certain that at the present stage of scientific development we are not in a position to derive 'theoretically' all the conditions which must be satisfied so that the six possible results of the game of dice will occur with equal frequency in a long series of throws. By 'theoretically' we mean a procedure which may make use of some results of experimental science but does not involve statistical experiments carried out with the apparatus whose probability we want to know, or with one similarly constructed.

The following proposition, although not an integral part of our new foundation of the theory of probability, is an essential element in my conception of statistical processes. The form of a distribution in a collective can be deduced only from a sufficiently long series of repeated observations, and this holds true for uniform as well as for all other distributions. The experiment may be carried out by using the system under consideration or one considered to be equivalent to it on the basis of appropriate observations. This proposition applies, in the first instance, to the distributions in the initial collectives with which all probability problems begin; it also applies to the distributions in the derived collectives if these are to be checked by observations.

According to our conception, the uniform distribution of probabilities is only a special case of the general distribution; this position is in sharp contrast to that of those epistemologists who uphold the so-called subjective theory of probability.

In the opinion of these authors, the probability which we ascribe to a certain event, i.e., to our assertion of its occurrence, depends exclusively on the degree of our knowledge; the assumption of equal chances for several events follows from our absolute lack of knowledge. I have already quoted the characteristically concise formulation of this principle due to E. Czuber, who said that we consider events to be equally probable if we have 'not the slightest knowledge of the conditions' under which each of them is going to occur. In an apparently more scientific form, this is the so-called 'Principle of Indifference'.

J. M. Keynes remarks, quite justly, that, by virtue of this principle, each proposition of whose correctness we know nothing, is endowed with a probability of 1/2, for the proposition and its contradictory proposition can be regarded as two equally likely cases. Thus, if we know nothing about the colour of the cover of a book and say that it is red, the probability of this assertion is 1/2. The same probabilities can also be asserted for the propositions that it is blue, yellow, or green, and consequently the sum of these probabilities is much larger than unity. Keynes[7] makes every effort to avoid this dangerous consequence of the subjective theory, but with little success. He gives a formal rule precluding the application of the Principle of Indifference to such a case, but he makes no suggestion as to what is to replace it. It does not occur to him to draw the simple conclusion that if we know nothing about a thing, we cannot say anything about its probability.

The curious mistake of the 'subjectivists' may, I think, be explained by the following example. If we know nothing about the stature of six men, we may presume that they are all of equal height. This application of the Principle of Indifference is also legitimate from the point of view of Keynes's rule. This presumption may be true or false; it can also be described as more or less probable, in the colloquial meaning of this word. In the same way we can presume that the six sides of a die, of whose properties we know nothing definite, have equal probabilities. This is, however, only a conjecture, and nothing more. Experiment may show that it is false, and the pair of dice used in our first lecture was an illustration of such a case. The peculiar approach of the subjectivists lies in the fact that they consider 'I *presume* that these cases are equally probable' to be equivalent to 'These cases *are* equally probable', since, for them, probability is only a subjective notion. Nobody, however, would assert that the above-mentioned six men are, in fact, equally tall, because the length of the body is something which can be measured objectively. If we were to maintain this difference between length and its probability, equal probabilities could in fact be 'deduced' from a lack of knowledge; we should, however, have just as much right to 'deduce' any other assumption concerning these probabilities, e.g., that they are proportional to the squares of the numbers 1 to 6, and this conjecture would be as permissible as any other.

I quite agree that most people, asked about the position of the centre of gravity of an unknown cube, will answer 'It probably lies at the centre'. This answer is due, not to their lack of knowledge concerning this particular cube, but to their actual knowledge of a great number of other cubes, which were all more or less 'true'. It would not be impossible to carry out a detailed psychological investigation into the foundations of our subjective probability estimations, but its relation to probability calculus is similar to that of the subjective feeling of temperature to scientific thermodynamics. Thermodynamics had its starting point in the subjective impressions of hot and cold. Its development begins, however, when an objective method of comparing temperatures by means of a column of mercury is substituted for the subjective estimate of the degree of warmth. Everyone knows that objective temperature measurements do not always confirm our subjective feeling, since our subjective estimate is often affected by influences of a psychological or physiological character. These discrepancies certainly do not impair the usefulness of physical thermodynamics, and nobody thinks of altering thermodynamics in order to make it agree with subjective impressions of hot and cold. I have previously pointed out that repeated observations and frequency determinations are the thermometers of probability theory.

BERTRAND'S PARADOX[8]

The attempts to justify, in various ways, the assumption of equally likely cases or, more generally, of a uniform distribution by having recourse to principles of symmetry or of indifference fails definitely in the treatment of the problems first considered by Bertrand, and later given the name of 'Bertrand's Paradox' by Poincaré. I shall show, by means of the simplest possible example, the insuperable difficulties which such problems present to every form of the classical theory of probability.

Consider the following simple problem: We are given a glass containing a mixture of water and wine. All that is known about the proportions of the liquids is that the mixture contains at least as much water as wine, and at most, twice as much water as wine. The range for our assumptions concerning the ratio of water to wine is thus the interval 1 to 2. Assuming that nothing more is known about the mixture, the indifference or symmetry principle or any other similar form of the classical theory tells us to assume that equal parts of this interval have equal probabilities. The probability of the ratio lying between 1 and 1.5 is thus 50%, and the other 50% corresponds to the probability of the range 1.5 to 2.

But there is an alternative method of treating the same problem. Instead of the ratio water/wine, we consider the inverse ratio, wine/water; this we know lies between 1/2 and 1. We are again told to assume that the two halves of the total interval, i.e., the intervals 1/2 to 3/4 and 3/4 to 1, have equal probabilities (50% each); yet, the wine/water ratio 3/4 is equal to the water/wine ratio 4/3. Thus, according to our second calculation, 50% probability corresponds to the water/wine range 1 to 4/3 and the remaining 50% to the range 4/3 to 2. According to the first calculation, the corresponding intervals were 1 to 3/2 and 3/2 to 2. The two results are obviously incompatible.

Similar contradictions may occur in all cases where the characteristic attributes (in our case the relative concentration) are expressed by continuous variables rather than by a discrete set of numbers (as in the case of a die or a lottery). I have already mentioned these so-called 'problems of geometrical probability', which owe their name to the geometrical origin of most of the older problems in this class. One of the oldest and best-known examples of geometrical probability is Buffon's needle problem (1733).[9] A number of parallel lines are drawn on the floor and a needle is dropped on it at random. The question is: What is the probability that the needle shall lie across one of the lines on the floor? The characteristic attribute of a single throw is the position of the needle in relation to the system of lines on the floor; it can be described by a set of numbers, called co-ordinates. Certain values of the co-ordinates correspond to the attribute 'crossing', other values to 'noncrossing'. The origin of possible contradictions in this case is exactly the same as in the preceding one. They arise from the fact that the results of the experiments can be described by means of several different sets of co-ordinates. The concentration of the mixture in the previous example could be described by the ratio wine/water as well as by the ratio water/wine. In the case of the needle, we can use Cartesian co-ordinates, polar co-ordinates, or other sets of co-ordinates. Equal probabilities for equal ranges in a certain co-ordinate system correspond, in general, to unequal probabilities for equal ranges in another possible co-ordinate system, and vice versa.

Any theory which starts from the notion of equal possibilities of a number of different cases, supposed to be known a priori, or derived by some kind of instinctive insight, must invariably fail when it comes to problems of this kind. Keynes, whom I have already mentioned as being one of the leading subjectivists, actually admits that in these cases several different assumptions are equally justifiable even though they lead to different conclusions. The point of view of the frequency theory is that in order to solve problems of this kind (as well as any other problems) the distribution in the initial collective must be given. The source of this knowledge and the special character of the distribution have nothing to do with probability calculus. In order to obtain results in an actual case which can be expected to be confirmed by experiment, the initial data must be taken from statistical observations. In the case of the water/wine mixture, it is perhaps difficult to give a reasonable definition of the collective involved; one would have to specify the actual procedure by which mixtures with different concentrations are obtained. In the case of the needle problem, the way in which the collective is formed is more or less clear. A needle is thrown repeatedly, by means of an arrangement whose details remain to be defined, and the distribution in this initial collective, which is formed by the series of throws, is characterized by a 'probability density', which may be given, in principle, in terms of any co-ordinate system. Once this density function has been determined by actual experiment, all further calculations must be based on it, and the final results are independent of the choice of co-ordinates, which are nothing but a tool. The problem belongs to the class of 'mixing' problems: all co-ordinate values corresponding to the crossing of a line by the needle are 'mixed' together and all the remaining values are similarly 'mixed'. It may be possible to choose co-ordinates such that the initial distribution is uniform in them. This may make the calculations easier; it is, however, of no importance. Some choices of co-ordinates may appear preferable from various points of view; none of them is indicated by an inherent necessity, though empirical conditions may influence our choice.

A SUGGESTED LINK BETWEEN THE CLASSICAL AND THE NEW DEFINITIONS OF PROBABILITY

As we have seen, the essential objections which can be raised against the classical definition of probability are twofold. On the one hand, the definition is much too narrow; it includes only a small part of the actual applications and omits those problems which are most important in practice, e.g., all those connected with insurance. On the other hand, the classical definition puts undue emphasis on the assumption of equally possible events in the initial collectives. This assumption fails in all those cases of 'geometrical' probability which were discussed in the last few paragraphs.

Nothing worthy of mention, as far as I know, has been brought forward to meet the second objection. I think that this objection is usually left unanswered through a lack of interest rather than on positive grounds. As far as the first objection is concerned, nearly everybody who has followed the traditional course in the theory of probability will reply that the classical theory provides a link connecting the two definitions of probability; that, owing to this link, the problems which were eliminated at the outset, such as those of life insurance, may be dealt with; and that the results are satisfactory, at least as far as practical applications are concerned. This link is supposedly found in the Law of Large Numbers, which was first suggested by Bernoulli and Poisson. (We have already mentioned it on a previous occasion.) By means of this law, it can be proved mathematically that probability values obtained as quotients of the number of favourable cases divided by the total number of equally possible cases, are, to a certain degree of approximation, equal to values obtained by the determination of relative frequencies in prolonged series of observations. Many authors have already pointed out the dangerous weakness of this link; nevertheless, it has been used again and again, owing to the absence of anything which could replace it.

We shall have to examine this point closely later on, owing to the general importance of the 'Law of Large Numbers' and the need of it in all practical applications. However, we postpone these delicate considerations for the moment. Our fourth lecture will deal exclusively with the various aspects of this famous law. Meanwhile, anticipating some of the results of that discussion, we state: The Law of Large Numbers, including its consequences, does not relieve us of the necessity of introducing probability as a limit of relative frequency. In fact, the basic law derived by Bernoulli and Poisson loses its main importance and actually its meaning if we do not adopt the frequency definition of probability. Only through hidden errors of argument and circular reasonings can we arrive at the idea that this theorem 'links' the frequency definition with the definition based on equally likely events.

SUMMARY OF OBJECTIONS TO THE CLASSICAL DEFINITION

The second part of this lecture will be a discussion of some new contributions to the foundation of the theory of probability; but before dealing with this, I would like to sum up briefly the objections I have raised against the classical definition of probability, based on the notion of equally likely events.

1. Since 'equally possible' is only another expression for 'equally probable', the classical 'definition' means, at best, a *reduction* of collectives with all kinds of distributions to collectives with uniform distributions.

2. Equally possible cases do not always exist, e.g., they are not present in the game with a biased die, or in life insurance. Strictly speaking, the propositions of the classical theory are therefore not applicable to these cases.

3. The statement that 'the six faces of an absolutely homogeneous cube have equal probabilities' is devoid of content, unless what we mean by 'equal probabilities' has previously been explained.

4. Perfect homogeneity, in the logical sense of this phrase, does not exist in practice. If the process of manufacture of a die is completely known, it is always possible to find aspects in which the different sides differ from each other.

5. The 'Principle of Indifference' and similar concepts are only circumlocutions of the classical theory. They avoid none of its difficulties.

6. In the case of a continuous distribution, the assumption of a 'uniform distribution' means something different in different

co-ordinate systems. No general prescription for selecting 'correct' co-ordinates can be given, and there can therefore be no general preference for one of the many possible uniform distributions.

7. The 'Law of Large Numbers', derived mathematically by Bernoulli and Poisson, provides no link between the definition of probability based on equally likely cases and the statistical results derived from long series of observations. It does not alter our postulate that the frequency definition is the starting point of the whole theory of probability. This last assertion will be elaborated on in the next lecture.

OBJECTIONS TO MY THEORY

Since my first publications which appeared in 1919, an intensive discussion of the foundations of the theory of probability has started and is still in progress. Those authors who had worked in this field for many years and had been successful in the solution of a number of special problems could hardly be expected to agree at once to a complete revision of the very foundations of their work. Apart from this older generation,[10] there is scarcely a modern mathematician who still adheres without reservation to the classical theory of probability. The majority have more or less accepted the frequency definition. A small group, whom I call 'nihilists', insist that basic definitions connecting probability theory with the empirical world are unnecessary. I will deal with this point of view at the end of this lecture.

Even among those who agree that the subject of probability calculus is frequencies and who think that this should find its expression in the definition of probability, there are still many differences of opinion. In the first place, there are some mathematicians who begin their course by defining probability as the limit of relative frequency, but do not adhere consistently to this point of view in their further developments. Instead, they revert to the old ways of the classical theory. The French textbook by Fréchet and Halbwachs (1924),[11] and that by the American mathematician Julian Coolidge (1925),[12] belong to this group.

A more recent work by Harald Cramér,[13] which seems to represent the prevalent trend among American and British statisticians, completely adopts the point of view of the frequency definition. Cramér rejects the definition based on equally possible cases as inadequate and firmly opposes the standpoint of the modern subjectivists which will be further discussed later on. However, Cramér omits giving a clear definition of probability and in no way explains or derives in a logical manner the elementary operations of probability calculus. The reason why he and authors of the same school of thought are able to proceed in this way is that, for all of them, the fundamental questions which arise from the *simple* problems of the theory of chance do not exist. If one's attention is focused on the mathematical difficulties of complicated problems it is easily possible to pass over the difficulties of the fundamentals. The same holds true in the case of pure mathematics: the mathematician who is concentrating on the solution of intricate problems need not concern himself with the proposition that a times b equals b times a. The significant difference is that in this field scientific discipline is much further advanced and it is therefore no longer customary to deal with the foundations in a few casual words.

Another small group of mathematicians is opposed to the definition of the collective as an infinite sequence of elements; they prefer to deal exclusively with frequencies in long, but finite, sequences, i.e., to avoid the use of limits. A larger group accepts my first postulate, viz., the existence of limiting values of relative frequencies, but finds difficulties with the second one, the postulate of randomness. Certain suggestions concerning the possible alteration of these conditions have been made. I propose to deal with these questions in turn in the following sections, including also a brief discussion of new developments in the subjective concept of probability.

FINITE COLLECTIVES

There is no doubt about the fact that the sequences of observations to which the theory of probability is applied in practice are all finite. In the same way, we apply in practice the mechanics of particles to the treatment of problems concerned with bodies of finite size which are not geometrical points. Nevertheless, nobody will deny the utility and theoretical importance of the abstraction underlying the concept of a material point, and this despite the fact that we now have theories of mechanics which are not based on the consideration of discrete points. On the other hand, abstractions that originally belonged to the mechanics of particles permeate far into the mechanics of finite bodies. We need not enter into details here.

It is doubtless possible to avoid the notion of infinite sequences in dealing with mass phenomena or repetitive events. The question is, what would be the results of such a method? I do not know of any argument for using infinite sequences, apart from the greater simplicity of this method, and I have never claimed for it any other advantages. In 1934, Johannes Blume[14] set himself the task of transforming my theory in such a way as to use only finite sequences of observations, especially in the fundamental definitions. His procedure is this: Instead of the postulate concerning the limits of the relative frequencies, he assumes the existence of certain fixed numbers determining the distribution of the collective, and postulates that the values of the actual relative frequencies should differ from these numbers by no more than a small positive magnitude ε. Assuming that ε is sufficiently small, it is possible to perform certain operations on these finite collectives, constantly remaining within the limits of an approximation defined by the magnitude ε. As far as this method actually goes, it amounts to nothing more than a circumscription of the concept of a limiting value, which may be quite useful for certain purposes. This has been stressed already by A. Kolmogoroff[15] in his review of Blume's work. The word 'limit' is in fact used in mathematics only as a concise way of making certain statements concerning small deviations. On the other hand, neither Blume nor other authors working in the same direction have so far been successful in describing in the language of the 'finite' theory all properties of a collective and all connexions between collectives, especially those relating to the principle of randomness. At the present time, therefore, I do not think that we can speak of the actual existence of a theory of probability based on finite collectives.[16]

Here I should like to insert an historical interpolation. The philosopher Theodor Fechner,[17] who had many-sided interests, created, under the name of 'Kollektivmasslehre', a kind of systematic description of finite sequences of observations, which he called 'finite populations' (Kollektivgegenstände). This work was edited by Lipps in 1897, after the death of the author. Fechner probably did not think of the possibility of arriving at a rational concept of probability from such an abstraction as his 'finite population', but his views have served, at least for me, as a stimulus in developing the new concept of probability.

Returning to our subject, I must defend myself most emphatically against the recurring misunderstanding that in our theory infinite sequences are always substituted for finite sequences of observations. This is of course false. In an example discussed at the end of the preceding lecture, we spoke of the group of twenty-four throws of a pair of dice. Such a group can serve as the subject of our theory, if it is assumed that it has been repeated, as a whole, an infinite number of times and in this way has become an element of a collective. This leads us to certain statements about probability that apply to a *finite* number of observations, in this example, twenty-four. Similarly, if we consider, for instance, the birth rate of boys in a hundred different towns, our theory shows what can be expected, on the average, in the case of this finite number ($n = 100$) of observations. There is no question of substituting an infinite sequence for each group of 100 observations. This point will be discussed in greater detail in the fifth lecture of this series, which will be concerned with the problems of statistics.

TESTING PROBABILITY STATEMENTS

The problem of formulating a theory of finite collectives, in the sense explained above, must be clearly distinguished from that of the actual interpretation of the results of our probability calculations. Since we consider that the sole purpose of a scientific theory is to provide a mental image of objectively observable phenomena, the only test of such a theory is the extent to which it applies to actual sequences of observations, and these are always finite.

On the other hand, I have mentioned on many occasions that all the results of our calculations lead to statements which apply only to infinite sequences. Even if the subject of our investigation is a sequence of observations of a certain given length, say 500 individual trials, we actually treat this whole group as one element of an infinite sequence. Consequently, the results apply only to the infinite repetition of sequences of 500 observations each. It might thus appear that our theory could never be tested experimentally.

This difficulty, however, is exactly the same as that which occurs in all applications of science. If, for instance, a physical or a chemical consideration leads us to the conclusion that the specific weight of a substance is 0.897, we may try to test the accuracy of this conclusion by direct weighing, or by some other physical experiment. However, the weight of only a finite volume of the substance can be determined in this way. The value of the specific weight, i.e., the limit of the ratio weight/volume for an infinitely small volume, remains uncertain just as the value of a probability derived from the relative frequency in a finite sequence of observations remains uncertain. One might even go so far as to say that specific weight does not exist at all, because the atomic theory of matter makes impossible the transition to the limit of an infinitely small homogeneous volume.

As a parallel to this difficulty we may consider, for instance, the fact that it is impossible to make an infinitely long sequence of throws with one and the same die, under unchanged conditions, because of the gradual wear of the die.

One could say that, after all, not all physical statements concern limits, for instance, that the indication of the weight of a certain finite volume of matter is likewise a physical statement. However, as soon as we begin to think about a really exact test of such a statement, we run into a number of conditions which cannot even be formulated in an exact way. For instance, the weighing has to be carried out under a known air pressure, and this notion of air pressure is in turn founded on the concept of a limit. An experienced physicist knows how to define conditions under which an experimental test can be considered as 'valid', but it is impossible to give a logically complete description of all these conditions in a form comparable, for instance, to that in which the premises of a mathematical proposition are stated. The assumption of the correctness of a theory is based, as H. Dubislav justly states, not so much on a logical conclusion (Schluss) as on a practical decision (Entschluss).

I quite agree with the view which Carl G. Hempel[18] put forward in his very clearly written article on 'The Content of Probability Statements'. According to Hempel, the results of a theory based on the notion of the infinite collective can be applied to finite sequences of observations in a way which is not logically definable, but is nevertheless sufficiently exact in practice. The relation of theory to observation is in this case essentially the same as in all other physical sciences.

Considerations of this kind are often described as inquiries into the 'problem of application'. It is, however, very definitely advisable to avoid the introduction of a 'problem of applicability', in addition to the two problems, the observations and their theory. There is no special theory, i.e., a system of propositions, deductions, proofs, etc., that deals with the question of how a scientific theory is to be applied to the actual observations. The connexion between the empirical world and theory is established in each case by the fundamental principles of the particular theory, which are usually called its axioms. This remark is of special importance to us because occasional attempts have been made to assign to the theory of probability the role of such a general 'application theory'. This conception fails at once when we realize that a new problem of application would arise in connexion with each single statement of the calculus of probability.

AN OBJECTION TO THE FIRST POSTULATE

The majority of mathematicians now agree that the concept of an infinite sequence of observations or attributes is an appropriate foundation for a rational theory of probability. A certain objection, resulting from a vague recollection of the classical theory, is raised, however, by many who hear for the first time the definition of probability as the limiting value of the relative frequency. I will discuss this objection briefly, although it does not stand close examination; it belongs logically to the problems which I am going to discuss in my next lecture dealing with the Laws of Large Numbers.

The objection[19] refers in fact to the text of the theorem of Bernoulli and Poisson which I have mentioned previously. According to this proposition, it is 'almost certain' that the relative frequency of even numbers in a very long sequence of throws with a correct die will lie near to the probability value $1/2$. Nevertheless, a certain small probability exists that this relative frequency will differ slightly from 0.5; it may be equal to 0.51, for instance, even if the sequence is a very long one. This is said to contradict the assumption that the limiting value of the relative frequency is exactly equal to 0.5.

In other words, so runs the objection, the frequency theory implies that, with a sufficient increase in the length of the sequence of observations, the difference between the observed relative frequency and the value 0.5 will certainly (and not *almost* certainly) become smaller than any given small fraction; there is no room for the deviation 0.01 from the value 0.50 occurring with a finite, although small, probability even in a sufficiently long sequence of observations.

This objection is based on nothing but an inexact wording and may be easily disposed of. The above-mentioned law does say something about the probability of a certain value of relative frequency occurring in a group of n experiments. We therefore have to know what probability means if we are to interpret the statement. According to our definition, the whole group of n consecutive throws has to be considered as one element in a collective, in the same way as this was done before with groups of four and of twenty-four throws. The attribute in the collective which we now consider is the frequency of the attribute 'even' in a group of n throws. Let us call this frequency x. It can have one of the $n + 1$ values, 0, $1/n$, $2/n$, . . . to $n/n = 1$. If 'even' appears m times in a series of n throws, the attribute is the fraction $x = m/n$. Each of these $n + 1$ different values of x has a certain probability. The probability that x has a value greater than 0.51 may be, for example, 0.00001. According to our theory, this means that if we repeat these sets of n throws an infinite number of times, we shall find that, on the average, 1 in 100,000 of these sets contains more than 51 % even results. The frequency which is considered in this example is that in a finite set of n casts and is obtained by the division of the m even numbers in the set by the fixed total number n of throws.

On the other hand, when defining the probability of 'even' we consider a relative frequency of a different kind. In fact, we consider the whole sequence of all experiments, without dividing it into sets of n, and count the number of even numbers from the beginning of the sequence. If N throws have been made altogether, and N_1 of them have given 'even' results, the quotient N_1/N is the frequency considered, and we assume that this fraction, in which both the denominator and the numerator increase indefinitely, tends to a constant limiting value. In our case this value would be 1/2. No immediate connexion exists between the two propositions of which one postulates the existence of a limiting value of the ratio N_1/N, for N tending to infinity, and the other states the occurrence of certain sets of the given fixed length n which exhibit an unusual value of the frequency m/n. There is therefore no contradiction between the two statements. The idea of such a contradiction could only arise from an incomplete and inexact formulation of the problem. One of the purposes of our next lecture will be to inquire more closely into the relation between these two statements, and we shall find not only that they are reconcilable but that the Law of Large Numbers acquires its proper sense and full importance only by being based on the frequency definition of probability.

OBJECTIONS TO THE CONDITION OF RANDOMNESS

I shall now consider the objections which have been raised to my second condition, that of *randomness*. Let us restate the problem. We consider an infinite sequence of zeros and ones, i.e., the successive outcomes of a simple alternative. We say that it possesses the property of randomness if the relative frequency of 1's (and therefore also that of 0's) tends to a certain limiting value which remains unchanged by the omission of a certain number of the elements and the construction of a new sequence from those which are left. The selection must be a so-called place selection, i.e., it must be made by means of a formula which states which elements in the original sequence are to be selected and retained and

87

which discarded. This formula must leave an infinite number of retained elements and it must not use the attributes of the selected elements, i.e., the fate of an element must not be affected by the value of its attribute.

Examples of place selection are: the selection of each third element in the sequence; the selection of each element whose place number, less 2, is the square of a prime number; or the selection of each number standing three places behind one whose attribute was 0.

The principle of randomness expresses a well-known property of games of chance, namely, the fact that the chances of winning or losing in a long series of games, e.g., of roulette, are independent of the system of gambling adopted. Betting on 'black' in each game gives the same result, in the long run, as doing so in every third game, or after 'black' has appeared five times in succession, and so on.

In my first publication in 1919 (see auto-bibliogr. note), I gave much space to the discussion of the concept of randomness. Among other propositions, I derived the following 'Theorem 5': 'A collective is completely determined by the distribution, i.e., by the (limits of the) relative frequencies for each attribute; it is however impossible to specify which elements have which attributes.' In the discussion of this proposition, I said further that 'the existence of a collective cannot be proved by means of the actual analytical construction of a collective in a way similar, for example, to the proof of existence of continuous but nowhere differentiable functions, a proof which consists in actually writing down such a function. In the case of the collective, we must be satisfied with its abstract "logical" existence. The proof of this "existence" is that it is possible to operate with the concept of a collective without contradictions arising.'

Today, I would perhaps express this thought in different words, but the essential point remains: A sequence of zeros and ones which satisfies the principle of randomness cannot be described by a formula or by a rule such as: 'Each element whose place number is divisible by 3 has the attribute 1; all the others the attribute 0'; or 'All elements with place numbers equal to squares of prime numbers plus 2 have the attribute 1, all others the attribute 0'; and so on. If a collective could be described by such a formula, then, using the same formula for a place selection, we could select a sequence consisting of 1's (or 0's) only. The relative frequency of the attribute 1 in this selected sequence would have the limiting value 1, i.e., a value different from that of the same attribute in the initial complete sequence.

It is to this consideration, namely, to the impossibility of explicitly describing the succession of attributes in a collective by means of a formula that critics of the randomness principle attach their arguments. Reduced to its simplest form, the objection which we shall have to discuss first asserts that sequences which conform to the condition of randomness do not exist. Here, 'nonexistent' is equivalent to 'incapable of representation by a formula or rule'.

A variant of this objection counters the joint use of the second with the first axiom, that of randomness with that of limiting values. The argument runs, roughly, as follows.

The existence or nonexistence of limiting values of the frequencies of numbers composing a sequence, say 1's and 0's, can be proved only if this sequence conforms to a rule or formula. Since, however, in a sequence fulfilling the condition of randomness the succession of attributes never conforms to a rule, it is meaningless to speak of limiting values in sequences of this kind.

RESTRICTED RANDOMNESS

One way to avoid all these difficulties would seem to consist in effectively restricting the postulate of randomness. Instead of requiring that the limiting value of the relative frequency remain unchanged for *every* place selection, one may consider only a predetermined definite group of place selections.

In the example which we discussed at the end of the second lecture, we made use of a frequently recurring, typical place selection. Starting with an infinite sequence of elements, we first selected the 1st, 5th, 9th, 13th, . . . elements; then the elements numbered 2, 6, 10, 14, . . .; following this, the numbers 3, 7, 11, 15, . . .; and finally 4, 8, 12, 16, . . . We assumed that in each of these partial sequences the limiting frequencies of the various attributes were the same as in the original sequence, and furthermore that the four partial sequences were 'independent' in the sense required for the operation of combination, i.e., that the limiting frequencies in the new sequences which are formed by combination and whose attributes are four-dimensional could be computed according to the simple rule of multiplication. The same reasoning holds true if instead of the value $n = 4$ we consider any other integral value for n, such as $n = 24$, or $n = 400$. A sequence of elements which has the above-described property for *every* n is today generally called a Bernoulli sequence. The American mathematician A. H. Copeland[20] and later on myself,[21] in a simpler way, have shown how it is actually possible to construct Bernoulli sequences. By following explicitly prescribed rules, one can form an infinite sequence of 0's and 1's which satisfies the above-stated conditions for every n.

Copeland has also shown that Bernoulli sequences have other interesting properties. If a partial sequence is formed out of those elements which follow a predetermined group of results, e.g., a group of five elements consisting of four 1's with a 0 in the middle, then in such a sequence the limiting frequency of the 1 (and of course also of the 0) will remain unchanged. We may therefore say that Bernoulli sequences are those without aftereffects. This property is called 'freedom from aftereffect'.

These facts seem to indicate that it might be sufficient to require that a collective should be of the Bernoulli type. Since it is explicitly possible to construct Bernoulli sequences, this restriction would dispose of all arguments against the existence of such collectives. Let us, however, consider what we would lose by thus restricting the condition of randomness.

Whereas we would undoubtedly be able to deal with questions of the type of the problem of the Chevalier de Méré, discussed in the preceding lecture, and would be able to proceed in the same way, there is, on the other hand, no doubt that a number of other meaningful questions would now remain unanswered. What happens, for instance, if a player decides, at the beginning, that he will consider only the first, second, third, fifth, seventh, eleventh, . . . casts of the die, that is to say, only those whose order number is a prime number? Will this change his chances of winning or not? Will the same rule of combination hold true in the sequence obtained through the place selection by prime numbers?

If, instead of restricting ourselves to Bernoulli sequences, we consider some differently defined class of sequences, we do not improve the state of affairs. In every case it will be possible to indicate place selections which will fall outside the framework of the class of sequences which we have selected. It is not possible to build a theory of probability on the assumption that the limiting values of the relative frequencies should remain unchanged only for a certain group of place selections, predetermined once and for all. All the same, we shall see that the consideration of sequences such as Bernoulli sequences and others, which satisfy conditions of restricted randomness, will prove valuable in solving certain questions in which we are interested.

MEANING OF THE CONDITION OF RANDOMNESS

In our theory of probability we have given first place to the proposition that in the sequence of observations under consideration the relative frequency of each attribute has a limiting value independent of any place selection. Let us review once more what we mean by this postulate. To be sure, it is not possible to prove it. Even if it were possible to form infinite series of observations, we would not be able to test any one of them for its insensitivity against *all* place selections, if for no other reason, because we are not in a position to enumerate *all* place selections. The axioms of science are not statements of facts. They are rules which single out the classes of problems to which they apply and determine how we are to proceed in the theoretical consideration of these problems. If we say in classical mechanics that the mass of a solid body remains unchanged in time, then all we mean is that, in every individual problem of mechanics concerned with solid bodies, it will be assumed that a definite positive number can be attributed to the body under consideration; this number will be called its mass and will figure as a constant in all calculations. Whether this is 'correct' or not can be tested only by checking whether the predictions concerning the behaviour of the body made on the basis of such calculations coincide with observations. Another reason for rejecting the axiom of a constant mass would be, of course, that it presented a logical contradiction with other assumptions. This, however, would merely imply that calculations based on all assumptions together would lead to mutually contradictory predictions.

Let us now see what kind of prescriptions follow from the axiom of randomness. After all that has been said in the first and second lectures, it can only be this: We agree to assume that in problems of probability calculus, that is, in deriving new collectives from known ones, the relative frequencies of the attributes remain unchanged whenever any of the sequences has been subjected to one or more place selections. We do not ask, at this moment, whether such an assumption is appropriate, i.e., whether it will lead us to useful results. All we ask now is whether this procedure may cause contradictions. This question can be answered clearly, as I shall show below. But first, I must insert some words of explanation introducing an important mathematical concept.

A quantity which cannot be expressed by a number, in the usual sense of the word, is said to be infinite. However, following Georg Cantor, the great founder of the theory of sets, modern mathematics

91

distinguishes between several kinds of infinity. I shall assume as known what is meant by the infinite sequence of natural numbers. If it is possible to establish a one-to-one correspondence between the elements of a given infinite set and the natural numbers, then we say that the set considered is enumerable or enumerably infinite. In other words, an infinite set is said to be enumerable whenever it is possible to number all its elements. The set of all numbers which represent squares of integers and also the set of all fractions having integers as numerators and denominators are enumerably infinite. On the other hand, the set of all numbers lying between two fixed limits, say, between 1 and 2, or the set of all points in a given interval are not enumerable. At least, it has not yet been possible to devise a theory of the set of points in an interval which would not use some other essential concept besides that of enumeration. The set of all points in an interval is said to be 'nonenumerable' or, more specifically, 'continuously infinite'. This distinction between enumerable and continuously infinite sets is of the greatest importance in many problems of mathematics. Using this concept, we will explain the present stage of our knowledge with respect to the consistency of the axiom of randomness.

CONSISTENCY OF THE RANDOMNESS AXIOM

During the last twenty years, a number of mathematicians have worked on this question. I name here in particular, K. Dörge,[22] A. H. Copeland,[23] A. Wald,[24] and W. Feller.[25] Although both the starting points and the aims of their respective investigations vary, all of them unequivocally bring out this *same* result: Given a sequence of attributes, the assumption that the limits of the relative frequencies of the various attributes *are insensitive to any finite or enumerably infinite set of place selections* cannot lead to a contradiction in a theory based on this assumption. It is not necessary to specify the type or properties of the place selections under consideration. It can be shown that, whatever enumerably infinite set of place selections is used, there exist sequences of attributes which satisfy the postulate of insensitivity. It can even be stated that 'almost all' (and this expression has a precise meaning which I cannot go into here) sequences of attributes have the required property. This last statement implies that collectives are in a sense 'the rule', whereas lawfully ordered sequences are 'the exception', which is not surprising from our point of view.

I know of no problem in probability in which a sequence of

attributes is subjected to more than an enumerably infinite number of place selections, and I do not know whether this is even possible. Rather, it might be in the spirit of modern logic to maintain that the total number of all the place selections *which can be indicated* is enumerable. Moreover, it has in no way been proved that if a problem should require the application of a continuously infinite number of place selections this would lead to a contradiction. This last question is still an open one.

But whatever the answer may be, from what we know so far, it is certain that the probability calculus, founded on the notion of the collective, will not lead to logical inconsistencies in any application of the theory known today. Therefore, whoever wishes to reject or to modify my theory cannot give as his reason that the theory is 'mathematically unsound'.

A PROBLEM OF TERMINOLOGY

I must now say a few words about another question, which is solely one of terminology. It has sometimes been said that a deficiency of my theory consists in the exclusion of certain purely mathematical problems connected with the existence of limiting values of relative frequencies in sequences of numbers defined by formulæ. It is not my intention to exclude anything. I have merely introduced a *new name*, that of a *collective*, for sequences satisfying the criterion of randomness. I think further that it is reasonable to use the word 'probability' only in connexion with the relative frequencies of attributes in sequences of this special kind. My purpose is to devise a uniform terminology for all investigations concerning the problems of games of chance and similar sequences of phenomena. It is open to everyone to use the term 'probability' with a more general meaning, e.g., to say that in going through the natural sequence of numbers the probability of encountering an even number is 1/2. It will, however, then be up to him to explain the difference existing, from his point of view, between, say, the natural sequence of integers and the sequence of the results of 'odd' and 'even' in a game of dice. This problem is not solved by a change in terminology.

Neither am I willing to concede that a theory is more general or superior because it is based on some notion of 'limited randomness', and therefore includes a greater variety of sequences. There still remains the essential difficulty of indicating the characteristics by which sequences such as those formed by the successive results of a game of chance differ from others. On the other hand, a probability theory which does not even try to define the boundaries of this special field, far from being superior to mine, fails, in my opinion, to fulfil the most legitimate demands.

I intend to show later, by certain examples, how sequences which do not possess the properties of collectives can be derived from collectives (in my sense) by means of operations which do not belong to the system of the four fundamental operations discussed above. In so far as sequences of this kind are of practical interest (e.g., certain so-called 'probability chains'), they belong within the framework of my theory; but I do not see any harm in denying the name 'probabilities' to the limiting values of the relative frequencies in such sequences. In my opinion, it is both convenient and useful to call these values simply 'limiting frequencies', or, as I have suggested earlier, to use a word such as 'chance'. Of course, there is no logical need for this cautious use of the word probability; it is quite possible that, once the frequency theory has been firmly established, more freedom can be allowed in the use of the terms.

OBJECTIONS TO THE FREQUENCY CONCEPT

As I have mentioned previously, the frequency theory of probability has today been accepted by almost all mathematicians interested in the calculus of probability or in statistics. This is usually expressed by the phrase that probability means an 'idealized frequency' in a long sequence of similar observations. I believe that by introducing the notion of the collective I have shown how this 'idealization' is obtained and how it leads to the usual propositions and operations of probability calculus.

On the other hand, there have been in the past and there still are a few authors who recommend applying the theory of probability in cases which in no way deal with frequencies and mass observations. To cite an older example: Eduard v. Hartmann,[26] in the introduction to his *Philosophy of the Unconscious* (1869), derives mathematical formulæ for the probability of natural events being due to spiritual causes, and finds it to be equal to 0.5904. I have earlier mentioned the economist John Maynard Keynes,[27] a persistent subjectivist. According to his opinion, probability ceases to be a trustworthy guide in life if the frequency concept is adopted. It seems to me that if somebody intends to marry and wants to find out 'scientifically' if his choice will probably be successful, then he can be helped, *perhaps*, by psychology, physiology, eugenics, or sociology, but surely not by a science which centres around the word 'probable'. The point of view of the geophysicist Harold Jeffreys[28] is similar to that of Keynes. In his book *Scientific Inference* (1931), he goes even further and says that any probability, in the widest sense of the word, can be expressed by a number. If, for example, a newborn child has seen only blue and red objects so far in his life, there exists for this child a numerical probability of the next colour being yellow; this probability, however, is not supposed to be determined in any way by statistical observations. Other arguments of the subjectivists have been presented earlier in connexion with the question of equally possible cases.

In recent years, the Keynes-Jeffrey point of view has found some support; efforts have been made to construct a rigorous system of subjective probability. Let us briefly describe these attempts.

THEORY OF THE PLAUSIBILITY OF STATEMENTS

In an interesting paper (1941), the mathematician G. Pólya[29] takes as his starting point the following historical fact. Referring to a proposition concerning an unproved property of integers, Euler stated that this proposition was 'probably correct', since it was valid for the numbers 1 to 40 as well as for the numbers 101 and 301. Even though such inductive reasoning is not otherwise customary in mathematics, or perhaps just because of this fact, Pólya considers this argument worthy of further investigation. He proposes that in such instances one might speak of 'plausibility' instead of probability. We are quite willing from our point of view to accept this terminological suggestion. Pólya arrives essentially at the following conclusions: (1) There are objective rules, i.e., rules accepted by all, on how to judge plausibility; e.g., if the number of known instances which support a proposition is increased, the plausibility increases; if an hypothesis on which the proposition could be founded is shown to be incorrect, the plausibility is decreased. (2) A numerically non-determinable figure, between 0 and 1, corresponds to every plausibility. (3) The formulæ of the calculus of probability are *qualitatively* applicable to plausibility considerations.

The first of the above conclusions, namely, that there are generally accepted rules for judging plausibility, will not be contended. What is meant by mathematical formulæ being qualitatively applicable is not quite clear to me. Perhaps this means that merely statements of inequalities and not of equalities can be made, though even that much would require that the plausibilities could be ordered in a sequence such as that of the real numbers. But my main objection to Pólya's plausibility theory is the following:

The plausibility of Euler's Theorem does not rest exclusively, or even essentially, on his forty-two particular instances. If it did, we might state equally well that all numbers of the decimal system could be represented by at most three digits, or that no number is the product of more than six prime numbers. The essential, or at least an essential, reason for the plausibility of Euler's theorem lies in the fact that it does not contradict any well-known and easily checked property of the integers. Moreover, if we pay attention to this theorem we do so because it was formulated by Euler and we know that he had a comprehensive knowledge of the theory of numbers. How are we to weigh these facts in judging the plausibility in question? Should we then count the number of properties which a theorem does not contradict? Would we have to conclude that plausibility will increase with every new property with which the theorem does not conflict?

As I have stated, Pólya does not attempt to express the plausibility of a statement by a definite number. Other authors are less reserved. R. Carnap,[30] who belonged to the Vienna Circle of Logical Positivism, now supports a theory of 'inductive logic' where he uses the expression 'probability 1' for the plausibility of a judgment, whereas the idealized frequency is called 'probability 2'. Both of these are said to follow the usual rules of probability calculus. In Carnap's opinion, the difference between Jeffrey's view and mine consists in the fact that one of us talks of 'probability 1' and the other of 'probability 2'. Within the framework of theory 1, Carnap formulates the following proposition: On the basis of today's meteorological data, the probability that it will rain tomorrow is 0.20. However, 'the value 0.20, in this statement, is not attributed to tomorrow's rain but to a definite logical relationship between the prediction of rain and the meteorological data. This relationship being a logical one . . . does not require any verification by observation of tomorrow's weather or any other observation.' Carnap does not state how the figure 0.20 is to be derived from the meteorological data. No meteorologist would fail to say that such a deduction is ultimately based on statistical experience. This, however, would bring us right back to probability 2. Carnap's theory would need to indicate how, by starting with propositions expressing the meteorological data, we arrive, by means of logical operations, at the figure 0.20 (or any other figure). His theory is, however, unable to show this.

The same unbridgeable gap exists in other systems which seek to define 'a purely logical notion of the plausibility of an hypothesis on the basis of given facts', using in an elaborate way the formal tools of symbolic logic and large doses of mathematics. C. G. Hempel and P. Oppenheim,[31] who attempted to do this, had to resort in the end to the admission of statistical observations as an essential basis, thus recognizing that mass phenomena and repetitive events are actually the subject of their theory. I certainly do not wish to contest the usefulness of logical investigations, but I do not see why one cannot admit to begin with that any numerical statements about a probability 1, about plausibility, degree of confirmation, etc., are actually statements about relative frequencies.

THE NIHILISTS

Finally, it is necessary to say a few words about those contemporary mathematicians who profess, more or less explicitly, that there is no need to give any definition or explanation of the notion of probability: What probability is, everybody knows who uses everyday language; and the task of the theory of probability is only to determine the exact values of these probabilities in different special cases. Such mathematicians completely misunderstand the meaning of exact science. I think that I have already said in the first lecture all that need be said about this question. It is essentially true that, historically, such a conception forms the starting point of scientific development. All theories arise primarily from the wish to find relations between certain notions whose meaning seems to be firmly established. In the course of such investigations, it is often found that not every notion for which the usual language has a word is an appropriate basis for theoretical deductions. In all fields in which science has worked for a sufficiently long time, a number of new artificial or theoretical concepts have been created. We know that this process is an essential part of scientific progress. Everywhere, from the most abstract parts of mathematics to the experimental physical sciences, in so far as they are treated theoretically, the exact definition of concepts is a necessary step which precedes the statement of propositions or goes parallel to it.

We may find an example in the modern development of physics. In the whole history of theoretical physics until the beginning of the present century, the notion of two simultaneous events occurring at two different points was considered to be self-evident and in no need of further explanation. Today, every physicist knows, as an essential

97

consequence of Einstein's special theory of relativity, that the notion of simultaneity requires a definition. A whole theory springs from this definition which is generally considered one of the most fruitful developments of modern physics. This theory must be simply non-existent for all who think that we know the meaning of simultaneity anyhow, i.e. 'from the usual sense of the word'.[32]

I think therefore that the refutation of those who consider every definition of probability to be superfluous can be left to follow its natural course. One reason for mentioning these 'nihilists' is the existence of certain intermediate opinions between their position and our point of view regarding the formation of concepts in an exact science. Some of these middle-of-the-road conceptions should not go unmentioned.

RESTRICTION TO ONE SINGLE INITIAL COLLECTIVE

A point of view typical of the attitude of many mathematicians is represented in A. Kolmogoroff's attractive and important book on the *Foundations of the Theory of Probability*.[33] To understand this point of view, consider for a moment the purely mathematical aspect of the content of a textbook on the theory of probability. We soon notice that a great many of the calculations are of one and the same type; namely, 'Given the distribution in a certain collective; to determine the probability corresponding to a certain part of the total set of attributes'; this 'part' of the so-called 'attribute space' or 'label space' is often determined in a complicated way; problems of this kind, which in our terminology belong to the class of 'mixing' problems, are sometimes very complicated. The following is an example:

The given collective consists of a combination of n simple alternatives, n being a very large number. The attribute of an element is thus a sequence of n symbols, which are, e.g., 0 or 1, 'red' or 'blue', etc. The probability of each combined result, i.e., each of the 2^n possible combinations of n symbols is known. We now consider another large number m, smaller than n, together with a variable number x, lying between m and n, and a given function $f(x)$, (e.g., the square root of x). One may now ask, what is the probability for the number of 1's among the first x symbols to be smaller than $f(x)$, for all x lying between m and n? This question obviously singles out a certain part of the 2^n possible combinations, a part depending only on the number m and the function $f(x)$, and we are seeking the sum of the probabilities of all attributes belonging to this group. This is a 'mixing' problem. The mathematical solution of such a problem can be a very difficult and complicated one, even if it consists, as in this case, in the application of one single fundamental operation to one given initial collective. In the literature, we find the solution of this problem for the special case of $f(x)$ proportional to the product $\sqrt{x} \log (\log x)$ with m and n both becoming infinitely large.

Let us return to the general problem in which we are interested. It is quite understandable that mathematicians who are engaged in the solution of difficult problems of a certain kind become inclined to define probability in such a way that the definition fits exactly this type of problem. This may be the origin of the view (which is in general not explicitly formulated), that the calculus of probability deals each time merely with one single collective, whose distribution is subjected to certain summations or integrations. This kind of theory would not need any foundation or 'axioms' other than the conditions restricting the admissible distributions and integrations. The axioms of this theory therefore consist in assumptions concerning the admissible distribution functions, the nature of the sub-sets of the attribute space for which probabilities can be defined, etc.

In the case of probability calculus, these basic mathematical investigations were carried out by Kolmogoroff. They form an essential part of a complete course on the theory of probability. They do not, however, constitute the foundations of probability but rather the foundations of the mathematical theory of distributions, a theory which is also used in other branches of science.

According to our point of view, such a system of axioms cannot take the place of our attempt to clarify and delimit the concept of probability. This becomes evident if we think of the simple case of the die or the coin where the above-indicated mathematical difficulties do not exist or rather where their solution is immediate without drawing on the mathematical theory of sets.[34]

Our presentation of the foundations of probability aims at clarifying precisely that side of the problem which is left aside in the formalist mathematical conception.

PROBABILITY AS PART OF THE THEORY OF SETS

By consistently developing a theory which deals with only one collective in each problem of probability and merely with one type of operation applied to this collective, we would eventually arrive at the conclusion that there is no theory of probability at all. All that is left of it then are certain mathematical problems of real functions and point sets which in turn can be considered as belonging to other well-known mathematical domains. 'From this point of view', to quote from one of the reviews of Kolmogoroff's book,[35] 'the theory of probability appears to lose its individual existence; it becomes a part of the theory of additive set functions'.

In the same manner, some mathematicians proclaimed that hydrodynamics does not exist as a separate science since it is nothing but a certain boundary problem of the theory of partial differential equations. Years ago, when Einstein's theory first became known among mathematicians, some of them said that electrodynamics is from now on a part of the theory of groups.

To a logical mind this identification of two things belonging to different categories, this confusion of task and tool is something quite unbearable. A mathematical investigation, difficult as it may be, and much space as it may occupy in the presentation of a physical theory, is never, and can never be, identical with the theory itself. Still less can a physical theory be a part of a mathematical domain. The interest of the scientist may be concentrated on the mathematical, i.e., the tautological, side of the problem; the physical assumptions on which the mathematical construction is based may be mentioned extremely casually, but the logical relation of the two must never be reversed.

Here is an analogy from another field: A state is not identical with its government; it is not a part of the governmental functions. In certain cases all the external signs of the existence of a state are the actions of its government; but the two must not be identified.

In the same sense probability theory can never become a part of the mathematical theory of sets. It remains a natural science, a theory of certain observable phenomena, which we have idealized in the concept of a collective. It makes use of certain propositions of the theory of sets, especially the theory of integration, to solve the mathematical problems arising from the definition of collectives. Neither can we concede the existence of a separate concept of probability based on the theory of sets, which is sometimes said to contradict the concept of probability based on the notion of relative frequency.

All that remains after our study of the modern formal development of this problem is the rather unimportant statement that the theory of probability does not require in its summations (or integrations) other mathematical implements besides those already existing in the general theory of sets.

DEVELOPMENT OF THE FREQUENCY THEORY

During the past decade, the frequency theory founded on the notion of the collective has given rise to a noteworthy development. This evolution seems most promising even though practically applicable formulations have so far not resulted from it. This new theory was founded in Germany (1936) by E. Tornier.[36] J. L. Doob[37] is today its chief proponent in America. I shall briefly explain its fundamental ideas, in so far as this is possible without presupposing familiarity with the theory of sets on the part of the reader.

At the outset, Tornier introduces in place of the 'collective' the concept of the 'experimental rule'. By that he means an infinite sequence of observations made according to a certain rule; for example, the consecutive results of a game of roulette. He expressly admits the possibility of the result of a certain observation depending on the preceding one or of other connexions. My theory is based on the assumption that all that happens to *one* given die, or to *one* given roulette wheel forms one infinite sequence of events. In Tornier's theory, however, a given experimental rule admits of an infinite number of infinite sequences as its 'realizations'. Let us, for instance, think of a game of 'heads and tails' with the possible results described by the figures 0 (heads) and 1 (tails). One realization of this game may be an infinite sequence of 0's, another a sequence of alternating 0's and 1's, in short, any infinite sequence consisting of these two numbers. The total of all possible realizations forms a set in the mathematical sense of the word; each group of realizations which have a certain characteristic in common is a partial set. If we assign the measure 1 to the total set, then the theory of sets teaches us how to attribute smaller numbers to the partial sets according to their frequencies; the sum of these numbers must be 1. In Tornier's theory, a given die, or rather the experimental rule referring to this die, is characterized by attributing to the partial sets of possible realizations certain measures as their probabilities. For instance, there may be a die such that the realizations containing more 1's than 6's predominate; for another die, sequences showing a certain regular alternation of results may occur frequently; and so on.

In Tornier's theory, there is not simply a probability of the 6 as such; there exists instead a probability of the 6 being, for instance, the result of the tenth cast, i.e., the relative frequency of the realizations which show a 6 on the tenth place. That means, of course, that the setup in Tornier's theory is much more general than that in my theory. His theory permits us, for instance, to stipulate that the probability of a 6 on the twentieth cast should be different from that on the tenth. It also leaves us free to make an arbitrary assumption concerning the probability of casting 1 in the eleventh trial after having cast 6 in the tenth one (this being the frequency of the group of realizations containing 6 in the tenth place and 1 in the eleventh place). Thus the multiplication rule does not follow from the fundamentals of this theory. Tornier's theory is also applicable to experimental rules whose results do not form collectives in the sense of my theory. To take into account the conditions which prevail in games of chance, it is necessary to make certain assumptions, e.g., that the multiplication rule holds, that the frequency of the realizations having a 6 in the nth place is independent of n, etc.

The greater generality of the Tornier-Doob theory is bought at the expense of a greatly complicated mathematical apparatus, but the logical structure of the system is perhaps more lucid and satisfactory. We will have to wait and see how the solutions of the *elementary* problems of probability calculus will be developed in the new system. This seems to me to be the test for judging the foundations of a theory.

It should be noted that in the American literature this development of the frequency theory is often referred to under the heading of 'Probability as a Measure of Sets'. I have earlier pointed out that probability can always be considered as a measure of a set even in the classical theory of equally likely cases. This is certainly not a speciality of the theory which we have just discussed, even though in it the principles of the theory of sets are used to a greater extent than in others.

SUMMARY AND CONCLUSION

I have said all that I intended to say on the problem of the foundations of the theory of probability and the discussion which has arisen around it, and I am now at the end of this argument. In an attempt to summarize the results, I may conveniently refer to the content of the last paragraphs. My position may be described under the following five points:

1. The calculus of probability, i.e., the theory of probabilities, in so far as they are numerically representable, is the theory of definite observable phenomena, repetitive or mass events. Examples are found in games of chance, population statistics, Brownian motion, etc. The word 'theory' is used here in the same way as when we call hydrodynamics, the 'theory of the flow of fluids', thermodynamics,

the 'theory of heat phenomena', or geometry, the 'theory of space phenomena'.

2. Each theory of this kind starts with a number of so-called axioms. In these axioms, use is made of general experience; they do not, however, state directly observable facts. They delineate or define the subject of the theory; all theorems are but deductions from the axioms, i.e., tautological transformations; besides this, to solve concrete problems by means of the theory, certain data have to be introduced to specify the particular problem.

3. The essentially new concept of our theory is the collective. Probabilities exist only in collectives and all problems of the theory of probability consist in deriving, according to certain rules, new collectives from the given ones, and calculating the distributions in these new collectives. This idea, which is a deliberate restriction of the calculus of probabilities to the investigation of relations between distributions, has not been clearly carried through in any of the former theories of probability.

4. The exact formulation of the necessary properties of a collective is of comparatively secondary importance and is capable of further modification. These properties are the existence of limiting values of relative frequencies, and randomness.

5. Recent investigations have shown that objections to the consistency of my theory are invalid. It is not possible to substitute for the general randomness requirement some postulate of randomness which is restricted to certain classes of place selections. The new set-up of Tornier and Doob constitutes a promising development of the frequency theory.

The Laws of Large Numbers[1]

AMONG the many difficult questions connected with the rational foundation of the theory of probability none has caused so much confusion as that concerning the real meaning of the so-called Law of Large Numbers, and especially its relation to the frequency theory of probability. Most authors waver between two positions: the definition of probability as the limiting value of relative frequency is alleged either to imply Poisson's Law[2] or to contradict it. In fact, neither is the case.

The plan of these lectures naturally includes a detailed discussion of this question. A restriction, however, is imposed on me by the fact that I do not expect from my audience any special mathematical knowledge; therefore I shall refrain from deductions of a mathematical kind. Nevertheless, I hope to be able to explain the essential points of the problem sufficiently clearly. We are going to discuss, besides the proposition which is usually called the Law of Large Numbers, its classical counterpart, often called the Second Law of Large Numbers, and we shall briefly indicate the extensions which these two laws have found in modern mathematics.

POISSON'S TWO DIFFERENT PROPOSITIONS

The ultimate cause of the confusion lies in Poisson's book itself. As we have already mentioned, he called two different propositions, which were discussed in two parts of his *Recherches sur la probabilité des jugements*, by the same name. Quite probably he held them to be really identical. In the introduction to his book he formulates the first of them in the following words: 'In many different fields, empirical phenomena appear to obey a certain general law, which can be called the Law of Large Numbers. This law states that the ratios of numbers derived from the observation of a very large number of similar events remain practically constant, provided that these events are governed partly by constant factors and partly by variable factors whose variations are irregular and do not cause a systematic change in a definite direction. Certain values of these relations are characteristic of each given kind of event. With the increase in length of the series of observations the ratios derived from such observations come nearer and nearer to these characteristic constants. They could be expected to reproduce them exactly if it were possible to make series of observations of an infinite length'.

These sentences, taken together with the discussion of a great number of practical examples which follows, make it quite clear that, in speaking of the Law of Large Numbers, Poisson meant here a generalization of empirical results. The ratios to which he refers in his proposition are obviously the relative frequencies with which certain events repeat themselves in a long series of observations, or the frequencies with which the different possible results of an experiment occur in a prolonged series of trials. If a certain result occurs m times in n trials, we call m/n its 'relative frequency'. The law formulated by Poisson in his introduction is thus identical with the first condition we imposed on a collective, namely, that the relative frequency of a certain event occurring in a sequence of observations approaches a limiting value as the sequence of observations is indefinitely continued. If, when speaking of the Law of Large Numbers, everybody meant only what Poisson meant by it in the introduction to his book, it would be correct to say that this law is the empirical basis of the definition of probability as the limiting value of relative frequency.

A large part of Poisson's book, however, is taken up by the derivation and discussion of a mathematical proposition, which the author also calls the Law of Large Numbers, and which is usually referred to either under this name or simply as 'Poisson's Law'. This proposition is a generalization of a theorem formulated earlier by Jacob Bernoulli.[3] The Bernoulli Theorem may be quoted as follows:

If an experiment, whose results are simple alternatives with the probability p for the positive result, is repeated n times, and if ε is an arbitrary small number, the probability that the number of positive results will be not smaller than n(p − ε), and not larger than n(p + ε), tends to 1 as n tends to infinity.

We may illustrate Bernoulli's proposition with a concrete example. In tossing a coin 100 times, we have a certain probability that the result 'heads' will occur at least 49, and at most 51 times. (Here the p of the theorem equals $1/2$, $n = 100$, $\varepsilon = 0.01$.) In casting the same coin 1000 times, the probability of the frequency of the result 'heads' being between 490 and 510, is larger (p and ε are the same as before, but $n = 1000$). The probability of this frequency falling in the range between 4900 and 5100 in 10,000 casts is still nearer to 1, and so on. Poisson's generalization of this proposition consists in discarding the condition that all casts must be carried out with the same or with identical coins. Instead, he allowed the possibility of using a different coin each time, p (in our case equal to $\frac{1}{2}$) now denoting the arithmetical mean of the n probabilities of the n coins.

A still more general and very simple formulation of this proposition was given by Tschebyscheff.[4] It applies to the case in which the experiment involved is not an alternative ('heads or tails'), but admits of a number of different results. For our discussion, however, of the fundamental meaning of Poisson's Law, it is quite sufficient to consider it in the special form which it takes in a simple game of '0 or 1'. The question is: What is the relation of this mathematical proposition, which we may briefly call Poisson's Theorem (or Bernoulli Theorem) to the empirical law formulated by Poisson in his introduction? Is it true that Poisson's Theorem is equivalent to this law? Is it correct to consider Poisson's Theorem as a theoretical deduction capable of an experimental test, and actually confirmed by it?

EQUALLY LIKELY EVENTS

To answer the above questions, we must begin by considering what Bernoulli and his successors understood by probability. Poisson's Theorem contains the word 'probability'; Poisson's empirical law does not mention it. To understand clearly the meaning of Poisson's Theorem in the classical theory, we must explicitly introduce into it the definition of probability used by Poisson himself.

We already know that the classical theory, in its concept of probability, did not take into account the frequency with which different events occur in long series of observations. Instead it declared, in a more formalist way, that 'probability is the ratio of the number of favourable cases to the total number of equally likely cases'. With an ordinary coin the two possible positions after the throw are the two 'equally likely' cases. One of them is favourable for the result 'heads'; thus 'heads' has the probability 1/2. This probability concept is the only one used in the derivation of Poisson's Theorem. To say that an event has a probability 'nearly 1' means, in the language of this theory, to stipulate that 'nearly all' the equally likely cases are favourable to the occurrence of this event.

If n throws are carried out with a coin, and n is a large number, the number of different possible results of this series of throws is a very large one. For instance, the first twenty throws, as well as the last thirty, may have the result 'heads', and all the remaining ones the result 'tails'; or the first ten throws may have the result 'tails', and the rest 'heads', and so on. With $n = 100$ there are 2^{100} (a 31-digit number) different possible outcomes of the game. If we assume that the probability p of throwing 'heads' is equal to $1/2$ for each single throw, then all these 2^{100} combinations of results must be considered as 'equally likely' cases. Let us assume that ε is taken as 0.01; Poisson's Theorem states that when n is a large number, by far the greatest part of the 2^n different results have the common property that the number of 'heads' contained in them differs from $n/2$ by not more than $n/100$. This is the content of the proposition derived by Poisson. It does not lead to any statement concerning the actual results of a prolonged series of experiments.

ARITHMETICAL EXPLANATION

In order to make this point still clearer, we shall now represent the results of throwing the coin by the figures 0 and 1, where 0 stands for the result 'heads' and 1 for the result 'tails'. Each game of 100 throws can be characterized by a 100-digit number, the digits being 0's and 1's. If we omit any zeros which precede the first 1 on the left-hand side, we obtain shorter numbers which can still be used to represent the corresponding sequence of experiments. We can now arrange all the numbers occurring in our system of results in a simple scheme, which begins as follows:

$$0, \ 1, \ 10, \ 11, \ 100, \ 101, \ 110, \ 111, \ 1000, \ 1001, \ \ldots$$

The scheme includes all numbers that can be expressed by 0's and 1's up to the number represented by a succession of 100 1's. As mentioned above, this sequence includes a total of 2^{100} numbers, i.e., about a million trillions.

The meaning of the notation introduced may be explained by the following example. The number 101 in the scheme corresponds to a result beginning with 97 zeros and ending with 1, 0, and again 1. If n were 1000 instead of 100, the scheme would begin with the same

107

numbers as above, but would be very much longer, containing 2^{1000} numbers. 0 would now mean a result composed of 1000 zeros, and 101 a result beginning with 997 zeros. Poisson's Theorem is then nothing but a statement concerning such systems of numbers.

The following facts are of a merely arithmetical nature and have nothing to do with repeated events or with probability in our sense.

If we consider the set of natural numbers represented by 0's and 1's up to 100 digits, the proportion of numbers containing 49, 50, or 51 zeros is found to be about 16%. If we extend the scheme to include numbers with up to 1000 figures, the proportion of those containing from 490 to 510 zeros is much higher, roughly 47%. Among the combinations of 0's and 1's containing up to 10,000 figures, there will be more than 95% containing from 4900 to 5100 zeros. In other words, for $n = 10,000$ at most 5% of all combinations are such that the proportion of 0's to 1's differs from $1/2$ by more than 1%, i.e., such that the number of 0's differs from 5000 by more than $0.01n = 100$. The concentration of the frequencies in the neighbourhood of the value $1/2$ becomes more and more pronounced with the increase in the length of the sequence of throws.

This arithmetical situation is expressed in the classical theory of probability by saying: In the first sequence the probability of the results 49 to 51 zeros is 0.16; in the second sequence the probability of the results 490 to 510 is 0.47; in the third sequence the results 4900 to 5100 have the probability 0.95. By assuming $\varepsilon = 0.01$ and $p = 1/2$, the theorem of Bernoulli and Poisson can be formulated as follows: Let us write down, in the order of their magnitudes, all 2^n numbers which can be written by means of 0's and 1's containing up to n figures. The proportion of numbers containing from $0.49n$ to $0.51n$ zeros increases steadily with an increase in n.

This proposition is purely arithmetical; it says something about certain numbers and their properties. The statement has nothing to do with the result of a single or repeated sequence of 1000 actual observations and says nothing about the distribution of 1's and 0's in such an experimental sequence. The proposition does not lead to any conclusions concerning empirical sequences of observations as long as we adopt a definition of probability which is concerned only with the relative number of favourable and unfavourable cases, and states nothing about the relation between probability and relative frequency.

The same considerations apply, in principle, to cases similar to that of tossing a coin. When we consider a game with a true die, we must replace the system of numbers composed of 1's and 0's by the system of numbers containing six different figures, i.e., the figures 1 to 6. The theorem states in this case that with an increase in the number n, the proportion of numbers containing about $n/6$ ones increases steadily, and finally approaches one.

Our conclusions can be summarized as follows: The mathematical deductions of Bernoulli, Poisson, and Tschebyscheff, based as they are on a definition of probability which has nothing to do with the frequency of occurrence of events in a sequence of observations, cannot be used for predictions relative to the results of such sequences. They have, therefore, no connexion whatsoever with the general empirical rule formulated by Poisson in the introduction to his book.

SUBSEQUENT FREQUENCY DEFINITION

How is it possible that Poisson himself considered his mathematical theorem as a confirmation of his empirical Law of Large Numbers?

Once this question has been asked, it is easy to answer. Poisson understood two different things by probability. At the beginning of his calculations, he meant by the probability 1/2 of the result 'heads' the ratio of the number of favourable cases to that of all equally possible cases. However, he interpreted the probability 'nearly 1' at the end of the calculation in a different sense. This value was supposed to mean that the corresponding event, the occurrence of between $0.49n$ and $0.51n$ heads in a game of n throws, must occur in nearly all games. This change of the meaning of a notion in the course of deduction is obviously not permissible. The exact point where the change takes place remains unspecified. Is the probability 0.16 of a series of 100 throws containing from 49 to 51 heads already to mean that 16% of all games in a long series must produce these results? Or is this interpretation only applicable to the probability 0.95 calculated for $n = 10,000$?

There is no possible answer to these questions. If one wishes to retain at all cost the classical definition of probability and at the same time to obtain Poisson's conclusion, then one must introduce an auxiliary hypothesis as a kind of *deus ex machina*.[5] This would have to run somewhat in this way: 'If a calculation gives a value not very different from 1 for the probability of an event, then this event takes place in nearly all repetitions of the corresponding experiment'. What else is this hypothesis but the frequency definition of probability in a restricted form? If a probability value 0.999 means that the corresponding event occurs 'nearly always', why not concede at once that the probability value 0.50 means that the event occurs in the long run in 50 cases out of 100?

It is then, of course, necessary to give a precise formulation of this assumption and to show that Poisson's Theorem can be derived on the basis of this new definition of probability. We shall see that this deduction of the theorem differs from the classical one in many ways. At any rate, the procedure of changing the meaning of a concept, without notice, between the beginning and the end of an argument is certainly not permissible.

We close this section with the statement that Poisson's Theorem can be brought into relation with Poisson's Law of Large Numbers only by adopting a frequency definition of probability in one form or another.

THE CONTENT OF POISSON'S THEOREM

It is natural to raise the following objection. If we wish to define the probability of the result 'heads' as the limiting value of its relative frequency, then we have to know in advance that such a limiting value exists. In other words, first of all we must admit the validity of Poisson's Law of Large Numbers. What then is the use of deducing, by complicated calculations, a theorem which apparently adds nothing to what has been assumed? The answer is that the propositions derived mathematically by Bernoulli, Poisson, and Tschebyscheff imply much more than a simple statement of the existence of limiting values of relative frequencies. Once the frequency definition of probability has been accepted and Poisson's Theorem restated in the terms of this theory, we find that it goes much further than the original Law of Large Numbers. Under the new conditions, the essential part of this theorem is the formulation of a definite statement concerning the succession of, say, the results 'heads' and 'tails' in an indefinitely long series of throws. It is easy to find sequences which obey Poisson's empirical Law of Large Numbers, but do not obey Poisson's Theorem. In the next section, we shall discuss a simple sequence of this kind. In it the relative frequency of the positive result has a limiting value 1/2, exactly as in tossing a coin; yet Poisson's Theorem does not apply to it. No mathematician who knows the properties of this sequence would consider it in connexion with probability. We shall use it her in order to show what additional condition is imposed by Poisson's Theorem on sequences which form the subject of the theory of probability, namely, those which we call collectives.

EXAMPLE OF A SEQUENCE TO WHICH POISSON'S THEOREM DOES NOT APPLY

Let us consider a table of square roots; this is a table containing the values of the square roots of the successive integers 1, 2, 3, 4, . . ., calculated to a certain number of decimals, say 7. We shall consider only the penultimate (sixth) figure after the decimal point; 'positive results' will be numbers which contain one of the figures 5, 6, 7, 8, or 9 in this place; all numbers containing one of the figures 0, 1, 2, 3, or 4 in this place will be 'negative results'. The whole table is transformed in this way into a sequence of 1's (positive results) and 0's (negative results) which alternate in an apparently irregular way. Theoretically, this sequence is an infinite one, although in practice all square-root tables end with some particular number. It is plausible and can be proved rigorously[6] that the limiting frequencies of both 0's and 1's in this sequence have the value 1/2. It is also possible to prove the more general proposition that the relative frequencies of the single figures 0, 1, . . . 9 are all equal to 1/10.

What we want to investigate is the possibility of applying Poisson's Theorem to this sequence. The theorem requires that if groups, each of n consecutive numbers, are formed from the infinite sequence of 0's and 1's, then, if n is sufficiently large, each group should contain about 50% zeros and 50% ones.

The beginning of the table seems to conform to this rule. To make this clear, we may consider each column of 30 entries in a specified table as a separate group, and then count the numbers of 'positive' and 'negative' results in each group. A simple calculation shows, however, that the state of affairs changes when we proceed to very large numbers which lie beyond the scope of the usual tables. According to a well-known formula, if a is much larger than 1, then the square root of an expression of the form $a^2 + 1$ is nearly equal to $a + \dfrac{1}{2a}$. Let us assume, for instance, that a is one million (10^6), and a^2 one billion (10^{12}). The square roots of the consecutive numbers a^2, $a^2 + 1$, $a^2 + 2$, . . . will differ by the very small amount $1/2 \; 10^{-6}$, i.e., by one half of a unit in the sixth decimal place. It is necessary to consider about ten consecutive entries to cause a change in the value of the square root by 0.000005, and so to change a 'positive' result in our sequence into a 'negative' one. In other words, in this part of the table our sequence of 0's and 1's contains regularly alternating groups of about ten consecutive 1's and ten consecutive

0's. The following section of an actual table illustrates this arrangement:

a^2	a	Case
$10^{12} + 1237$	$10^6 + 0.0006185$	1
$+ 1238$	6190	1
$+ 1239$	6195	1
$+ 1240$	6200	0
$+ 1241$	6205	0
$+ 1242$	6210	0
. . .	. . .	. . .
$+ 1249$	6245	0
$+ 1250$	6250	1
. . .	. . .	. . .

Further along in the table, e.g., in the region of $a = 100$ millions, the run of consecutive 0's and 1's will be of the average length 1000.

We see that the structure of the sequence of 1's and 0's derived from the table of square roots is radically different from the structure of sequences such as that derived from tossing a coin. The randomness described by Poisson's Theorem apparently exists only at the beginning of the table. Further on, the runs of identical 'results' are slowly but steadily increasing in length. It can be easily seen that this sequence does not obey the Poisson Theorem. Let us assume a large, but finite number n as the length of a group, say $n = 500$. By taking enough terms in our sequence (e.g., to $a = 100$ millions), we come into a region where the average length of the runs is much greater than n. In this region nearly all groups of 500 items will consist either of zeros only, or of ones only, whereas Poisson's Theorem requires them to consist of, roughly, 50% of each. The limiting values 1/2 for the frequencies of the results '1' and '0' are due in this case to the equal lengths of runs of both kinds. In a true game of chance, however, the approach to the limiting values is brought about by an equalization of the relative frequencies of 0's and 1's in nearly every group of sufficient length.

BERNOULLI AND NON-BERNOULLI SEQUENCES

What was shown here for the table of square roots holds true, in the same or in a similar way, for tables of many other functions, e.g., for powers of any order, etc. It does not hold for the table of logarithms, which Henri Poincaré discussed as an example. Poincaré failed to notice that in this table the frequencies of 0's and 1's fluctuate permanently and do not tend toward limiting values.

The result that is important for us is that there exist infinite sequences of 0's and 1's such that the relative frequencies of both these attributes tend toward definite limiting values but for which Bernoulli's theorem does not hold true.

Of course, the sequences which we have considered here do not satisfy the condition of randomness since each of them is defined by a mathematical formula. We may then ask whether randomness in our sense (which, together with the existence of the limit of relative frequencies forms a *sufficient* condition for the validity of Bernoulli's theorem) is a *necessary* prerequisite for the validity of this theorem. This is not the case. It is not too difficult to construct mathematical formulæ which define 'Bernoulli sequences', i.e., sequences where a stabilization of frequencies will occur in sufficiently long subsequences. Hence, the state of affairs is the following:

For *arbitrary* sequences satisfying the first axiom the Bernoulli Theorem need not hold true. It is not necessary to require complete randomness in order to prove the Bernoulli theorem. In other words, the theorem of Bernoulli is a consequence of the assumption of randomness but it cannot be used as a substitute for the randomness requirement. One could, for instance, indicate sequences of numbers which would satisfy the Bernoulli theorem but in which the chance of an attribute could be changed by a place selection of the prime-number type. The Bernoulli-Poisson-Tchebyscheff Theorem expresses only a special type of randomness. If we call those sequences for which this theorem holds 'Bernoulli-sequences', we can say that: All collectives are Bernoulli-sequences but not all Bernoulli-sequences are collectives.

DERIVATION OF THE BERNOULLI-POISSON THEOREM

It should thus be clear that, once the frequency definition of probability has been adopted, the theorem derived by Poisson in the fourth chapter of his book contributes important information regarding the distribution of the results in a long sequence of experiments. We have also seen that there is no logical way of obtaining this information starting with the classical definition of probability. The only way is to define probability from the beginning as the limiting value of relative frequencies and then to apply, in an appropriate manner, the operations of selection, combination, and mixing.

113

Let us consider the simplest case, in which an infinite sequence of experiments is made for the same simple alternative with the probability p of success (special case of the Bernoulli Theorem). By applying n selections followed by n combinations, we form a new collective just as we did when solving the problem of de Méré, in the second lecture (p. 00). Each element of this new collective consists of a group of n trials. The attribute of each such element is therefore given by n numbers (0's and 1's). Then, by the operation of mixing, we return to a one-dimensional attribute: This is done by considering now as the result of the combined experiment only the *number n_1 of* 1*'s* in the group of n trials, regardless of the arrangement of 0's and 1's in the group. In this way we obtain a probability of the occurrence of n_1 ones in a group of n trials. We call this probability $w(n_1; p)$, since it is dependent on both n_1 and p; it is given by the formula:

$$w(n_1; p) = \binom{n}{n_1} p^{n_1} (1 - p)^{n - n_1}.$$

Here the symbol $\binom{n}{n_1}$ stands for a known integer, dependent on n and n_1. If we add all the $w(n_1; p)$ for all those n_1 which fall between $n(p - \varepsilon)$ and $n(p + \varepsilon)$, where ε is an arbitrarily chosen, small positive magnitude, we obtain the probability P that the relative frequency of 1's in a group of n experiments lies between $(p - \varepsilon)$ and $(p + \varepsilon)$. Studying the resulting formula, we arrive at the conclusion that, however small the value of ε, the probability P tends towards unity as n increases indefinitely. *We have thus proved the Bernoulli Theorem.*

Today we know simpler and more far-reaching methods of arriving at this result. We can also include the more general case where the n observations are not derived from the same collective (case of Poisson), or where they are not simple alternatives (case of Tchebyscheff). These generalizations hold no further interest for us here since we are concerned only with the logical aspect of the deduction.

In closing, we note that by amending the usual understanding of the 'First Law of Large Numbers' we are in no way belittling the great achievement of the pioneers of probability calculus. This will be realized by anyone who is at all familiar with the history of mathematics. The founders of analysis, Bernoulli, Euler, Laplace, have given us a large treasure of propositions on integration, series, and similar subjects, which were properly derived and correctly fitted into logical systems only centuries later. In many cases, all that was needed was greater precision with respect to the passage to the limit.

To provide this greater precision in the case of the fundamentals of probability calculus was the very task which we had set ourselves.

SUMMARY

(1) The proposition introduced by Poisson in the introduction to his book as the Law of Large Numbers is a statement of the empirical fact that the relative frequencies of certain events or attributes in indefinitely prolonged sequences of observations tend to constant limiting values. This statement is postulated explicitly in our first condition for a collective.

(2) The mathematical proposition derived by Poisson in the fourth chapter of his book, which he also called 'The Law of Large Numbers', says nothing about the course of a concrete sequence of observations. As long as we use the classical definition of probability, this theorem is nothing but a statement of certain purely arithmetical regularities, and bears no relation to the empirical law explained in (1).

(3) The Poisson Theorem (see (2)) obtains a new meaning if we agree to define probability as the limiting value of relative frequency, a definition suggested by the above empirical law. If, however, we adopt this definition, Poisson's Theorem goes considerably further than the empirical law; it characterizes the way in which different attributes alternate in empirical sequences.

(4) The content of Poisson's Theorem is in fact that a certain equalization or stabilization of relative frequencies occurs already within nearly all sufficiently long sub-groups of the infinite sequence of elements. This is not implied by assuming only the existence of limiting values of relative frequencies. In fact, as was shown in an example of a sequence of 0's and 1's, the relative frequencies may tend to definite limits, e.g., 1/2, but the runs of 0's and 1's may gradually and regularly increase in length so that eventually, however large n may be, most groups of n consecutive elements will consist of 0's or of 1's only. Thus, in most of these groups there will be no 'equalization' or 'stabilization' of the relative frequencies; obviously, such a sequence does not satisfy the criteria of randomness.

(5) The correct derivation of the Poisson Theorem based on the frequency definition of probability requires not only the assumption of the existence of limiting values but also that of complete randomness of the results. This condition is formulated in our second postulate imposed on collectives.

After things have been clarified in this way, all that remains to be done is to decide a question of terminology: which proposition shall be called the 'Law of Large Numbers'? My suggestion is that this name should be reserved for the theorem of Bernoulli and Poisson. The empirical law formulated by Poisson in the introduction to his book can conveniently be called the axiom of the existence of limiting frequencies.

If the probability of the occurrence of an attribute within a given collective has a value near to unity, we may express this fact by saying that 'there is a great certainty' or 'we are almost certain' that this event will occur on *one* specific trial. This way of expressing ourselves is not too reprehensible so long as we realize that it is only an abbreviation and that its real meaning is that the event occurs almost always in an infinitely long sequence of observations. If we apply this terminology in connexion with the Bernoulli Theorem, and if we say in addition that a number which, for small ε, lies between $n(p - \varepsilon)$ and $n(p + \varepsilon)$ is 'almost equal to np', we arrive at the following imprecise formulation: If a trial with probability p for 'success' is performed again and again, and if n is a large number, it is to be expected with great certainty that in one particular sequence of n trials the event will occur approximately np times. This formulation leads us to ask whether a certain converse of this proposition might not hold true, namely the following: If in one set of n observations, n being a large number, the 'event' has occurred n_1 times, may we then inversely 'expect with great certainty' that the basic probability p of the event is approximately equal to the ratio n_1/n? We shall see that under certain conditions this is actually the case, and that this so-called Bayes's[7] Theorem represents a Second Law of Large Numbers which, under certain assumptions, can be proved mathematically.

If we use an extremely sloppy terminology, both laws can be said to coincide in one statement· The probability of an event and the relative frequency of its occurrence, in a long sequence of trials, are about equal. If, however, we use our words with precision—and to do so is one of the principal tasks of these lectures—we shall find that the original proposition and its converse have very different meanings. *Now*, we are speaking about a probability that p will assume certain values or will lie in a given interval, whereas in the original instance p was a given, unchanging number. I shall presently construct the collective for Bayes's problem, but first I wish to insert a remark.

Those who think that probability can be defined independently of the frequency of occurrence of an attribute in a sequence of experiments believe that the above-mentioned proposition, whereby probability and frequency roughly coincide in a long run of observations, constitutes a 'bridge' between what actually happens and the concept of probability introduced by them. However, we know that this is a delusion. From the definition of probability as the ratio of favourable to equally likely cases, no logical reasoning will lead to the propositions discussed above—neither to the original Bernoulli-Poisson statement nor to Bayes's converse of it. All that we can logically deduce from this premise is propositions concerning such ratios. A gap remains: the manner in which it is to be crossed is arbitrary and logically not justifiable.

BAYES'S PROBLEM

An easy way to understand Bayes's problem is to consider a game of dice in combination with a lottery. Let us imagine an urn filled with a very large number of small cubes or similar bodies, which we are going to call stones. Each stone has six flat sides, numbered 1 to 6, on any of which it can fall when thrown out of a dice box. Each time a stone is drawn from the urn, it is placed in a box and then turned out, and the result of the throw is noted. The stones are not all equal; some of them may be 'true', with a probability $1/6$ for each side, others biased, with the single probabilities differing more or less widely from the value $1/6$. The actual values of the six probabilities can be determined by a sufficiently long sequence of throws made with each stone. We shall consider the probability of casting 6, and denote it by p. Thus each stone has a characteristic value of p.

We now consider a collective whose elements are as follows: A stone is drawn from the urn, placed in the dice box, and thrown n times in succession. The result, or attribute, of this total experiment is composed of, on the one hand, the p-value of the stone drawn from the urn, and, on the other hand, the ratio n_1/n, where n_1 is the number of casts which have produced the result 6. This is a two-dimensional collective, with each element characterized by a pair of numbers, p and n_1/n.

The distribution in this collective, or the probabilities of all possible combinations of the numbers p and n_1/n, must be calculated by the combination rule. This involves the multiplication of two

117

factors: the first of them is the probability $v(p)$ of drawing a stone with a certain value of p from the urn; the second is the probability that an event with probability p will occur n_1 times in a series of n trials. The elementary theory of probability teaches us how to calculate this second factor from given values of p, n_1, and n. This probability has, in fact, been derived (see p. 114) and denoted by $w(n_1; p)$. The probability of drawing a stone with the attribute p and of obtaining with it n_1 6's in n throws is, according to the multiplication rule, $v(p)w(n_1; p)$. Assuming that n is a constant, we can consider this product as a function of p and n_1/n, and denote it by $f(p, n_1/n)$.

To arrive at Bayes's problem, we must perform one more operation on our collective, namely, a partition. Let us recollect what we mean by this. In a simple game of dice a typical problem of partition is the determination of the probability of the result 2, if it is already known that the result is an even number. The collective is divided into two parts, and only that corresponding to 'even' results is considered.

In the case which we are now discussing, the following partition takes place. We know already that 6 appeared n_1 times in n casts. What is the probability of this result 'being due' to a stone with a certain given value of p? The part of the collective to be considered in the calculation of this probability is that formed by the sequence of casts with a certain value of n_1/n, e.g., $n_1/n = a$. According to the division rule derived in the second lecture, the probability in question, the final probability, is calculated by dividing the probability $f(p,a)$ by the sum of the probabilities of all results retained after partition. The summation must be extended over all values of p, while a is constant. We will denote the result of this summation by $F(a)$, since it is a function of a only. The final result of the calculation is the quotient $f(p,a)/F(a)$, i.e., a function of both p and a. We may call it, in the usual terminology, the 'a posteriori probability' of a certain value of p, corresponding to a given value of $n_1/n = a$.

INITIAL AND INFERRED PROBABILITY

The expression 'a posteriori probability' for the ratio $f(p,a)/F(a)$ refers, of course, to the fact that there is still another probability of p in our problem, namely, $v(p)$. The latter is the probability that a stone with a given value of p will be drawn from the urn. It is called the 'a priori probability' of p. Even though we feel that it does not matter what words are used so long as their meaning is clear, it does

118

seem preferable to use expressions which are not weighed down by metaphysical associations. We have already proceeded in this way in our second lecture, when we developed the operation of partition.

The quantity $v(p)$ has the following meaning: Instead of considering the whole experiment, let us fix our attention only on the repetitive drawing of a stone out of the urn, without being interested in what happens later to this stone. The relative frequency with which a stone with a given value of p appears in these drawings has the limiting value $v(p)$. This value may well be called the *initial* or *original* probability of p. This probability is quite independent of the observations made subsequently on a stone during the n times it is tossed out of the dice box. Next, let us consider the whole experiment, concentrating only on those instances when, in n throws, a stone which may have any value of p showed the 6 with a relative frequency $n_1/n = a$. Among the stones thus selected, those with the specified value of p will appear with a relative frequency different from $v(p)$. The limiting value of this frequency, in an infinitely continued sequence of observations, will be $f(p,a)/F(a)$. This limiting value is the probability which we infer from the observation of the n_1 successes in n throws. We shall therefore call it the *probability of inference* or *inferred probability* of p. A numerical example will clarify this point.

Suppose our urn contains nine kinds of stones, such that the possible values of the probability p for the occurrence of a 6 are 0.1, 0.2, 0.3, . . ., 0.9. The stones are considered alike in outward appearance. They could, for instance, be regular icosahedra showing the 6 on 2, or 4, or 6, . . ., or 18 of their 20 sides with corresponding values of p equal to $2/20 = 0.1$, $4/20 = 0.2$, . . ., $18/20 = 0.9$. If there is an equal number of each kind in the urn, we can assume that the probability of drawing a stone from any of the nine categories is the same. Therefore, in this case, $v(0.1) = v(0.2) = . . . = v(0.9) = 1/9$. This gives us our initial distribution $v(p)$.

We now decide to consider the case $n = 5$, $n_1 = 3$, i.e., we cast each stone five times and note only those instances in which three of the five throws have shown a 6. The probability of obtaining three successes out of five trials with a given experimental object can be determined according to the formula shown on p. 114, and is:

$$w(3;p) = 10p^3(1 - p)^2.$$

In fact, if $n = 5$ and $n_1 = 3$, the term $\binom{n}{n_1}$ has the value 10. We

119

can now calculate the product $f(p,a)$, where $a = n_1/n = 3/5$, for each of the nine values of p. It is given by the formula:

$$f(p,3/5) = v(p)w(3;p) = \tfrac{1}{9} . 10p^3(1 - p)^2.$$

This gives us the following easily computed results: For $p = 0.1$, the corresponding value of f is 0.0009; for $p = 0.2$, $f = 0.0057$; . . .; for $p = 0.6$, $f = 0.0384$; and so on. The sum of all nine f-values amounts to $F = 0.1852$. Consequently, the probability of inference which we are seeking will be $0.0009/0.1852 = 0.005$ for $p = 0.1$; for $p = 0.2$ we obtain $0.0057/0.1852 = 0.031$; . . . for $p = 0.6$ it becomes $0.0384/0.1852 = 0.21$; and the sum of nine values will be equal to 1. We thus have shown the following: If we know nothing about a stone except that it was drawn from the above-described urn, the probability that it will belong to one of the nine different categories is equal to $1/9 = 0.11$. If, however, we already know that this stone has shown three successes in five casts, then the probability that it belongs to the category of $p = 0.1$ or $p = 0.2$ becomes much smaller than 0.11, namely, 0.005 or 0.031 respectively, whereas the probability that it is a stone for which $p = 0.6$ increases from 0.11 to 0.21. Everyone will consider this result to be understandable and reasonable. The fact that a stone has had a history of 60% success within a short experimental run increases the probability that it is a stone of the category of $p = 0.6$ and lessens its probability for a very different value of p such as 0.1 or 0.2. Other values of the probability of inference in this case are: 0.08 for $p = 0.3$; 0.19 for both $p = 0.5$ and $p = 0.7$ (taking two decimals); and finally 0.05 for $p = 0.9$. The sum of the three probabilities corresponding to the three values $p = 0.5$, $p = 0.6$, and $p = 0.7$ is 0.59, whereas the total probability corresponding to the other six p-values together is only 0.41.

LONGER SEQUENCES OF TRIALS

Let us consider the same example as above, but with a certain modification. We have again nine types of objects in equal proportions in our urn. Thus, the initial probability for each category is again $v = 1/9 = 0.11$ for each of the nine types. Once again, we 'partition off' those instances when casts with a stone have shown success with a frequency of 60%. However, this time we will assume that the total number of casts is no longer 5 but 500, and, correspondingly, that the number n_1 of required successes is changed from 3 to 300. What can we now infer with regard to the probabilities of the nine values of p?

120

The same formula as used above will answer our question, even though the process of computation is now more complicated, owing to the higher powers of p and $(1 - p)$, and to the higher value of $\binom{n}{n_1}$. The result is as follows: The probability of inference for $p = 0.6$ is now 0.99995, while the figure which corresponds to the sum of the three neighbouring p-values 0.5, 0.6, and 0.7 differs from unity only in the 17th decimal place. We can therefore say that it is 'almost certain' that an object which has had 60% 'success' in 500 trials has a probability p of success equal or almost equal to 0.60. This result is not surprising. It shows that inference based on a long series of experiments is far more effective than that based on a short one. If we consider all the stones in the urn, the same probability corresponds to each of the nine categories, whereas by considering only those stones which showed 60% success in 500 trials, we find, practically speaking, no probability corresponding to values of p smaller than 0.5 and larger than 0.7. This result stands out even more markedly as we increase the number n of observations still further. This feature constitutes the main content of Bayes's Theorem. Let us state it, for the moment, in a preliminary version:

If an object, arbitrarily drawn out of an urn, is subjected to a large number n of observations showing a frequency of success $n_1/n = a$, it is highly probable that the probability of success p of this object is equal or nearly equal to the observed relative frequency a. Stated a little more precisely, this probability tends to unity as n increases indefinitely.

We have to clarify further the meaning of the words 'equal or nearly equal to a'. If we assume that there is merely a finite number of different stones with corresponding different, discrete values of the probability p of success, then the observed frequency a of success will, in general, not coincide with any of those values of p. There will, however, be two values immediately adjoining the observed value of a which are equal to some of the possible p-values and we can apply the statement of Bayes's Theorem to these two. If, as in our case, the observed value of a does coincide with one of the possible values of p, then this one alone, or together with the two immediately adjoining values of p, is to be included in the statement of Bayes's proposition.

The mathematician will often prefer to stipulate that the probability of success p is not restricted in advance to certain discrete fractions but that it can assume any value in the whole interval between 0 and 1, or in a given part of this interval. In that case, the initial probability $v(p)$ is given as a probability density. Bayes's Theorem then considers p along a short interval extending from $a - \varepsilon$ to $a + \varepsilon$, where ε is an arbitrarily small number. We can then state the following: *If an object picked at random has shown a frequency of success a, in a long sequence of experiments, then the probability P that the probability p of this object lies between a − 'ε and a + ε will approach unity more and more closely as the number n of experiments is more and more increased.* The number ε can be arbitrarily small, so long as the interval from $a - \varepsilon$ to $a + \varepsilon$ is large enough to include, in the case of preassigned discrete p-values, at least two possible values of p.

INDEPENDENCE OF THE INITIAL DISTRIBUTION

We still have to add a very essential remark in order to bring into proper focus the actual content and the great importance of Bayes's proposition. We have seen that the probability of inference depends on two groups of data: (1) the initial probability $v(p)$; (2) the observed results of n experiments from which the inference is drawn. In our first example, we assumed that the nine different types of stones, with $p = 0.1$ to 0.9, were contained in equal numbers in the urn, so that the value of each of the nine probabilities $v(p)$ was equal to $1/9$. Let us now assume that there are in the urn one stone of the first category, two of the second, . . ., and nine of the ninth category. The total content will now be 45 stones (being the sum of the numbers 1 to 9), and the values of the probabilities will be: $v(0.1) = 1/45$, $v(0.2) = 2/45$, . . ., $v(0.9) = 9/45$. The probability of inference can again be computed according to the formula on p. 000, substituting, in place of the previous factor $1/9$, the new values $1/45$, $2/45$, . . ., $9/45$, respectively. We now obtain the following results from our calculations: The probability of inference for $p = 0.1$ is now 0.001 (compared with 0.005 before); for $p = 0.2$, it is now 0.011 (0.031 before); for $p = 0.5, 0.6, 0.7$, we get 0.16, 0.22, 0.23 (0.19, 0.21, 0.19 before), and for $p = 0.9$, we find 0.07 (0.05 before). As was to be expected, the new values are markedly different from the earlier ones. If we compare the inferred with the initial probabilities we find, however, again that the numerical results are higher for values of p close to 0.6, and lower for values further away from 0.6.

Let us now consider the same distribution of the initial probabilities but assume that the number n of experiments is 500. We then arrive at the very remarkable fact that now there is no noticeable change in the inferred probabilities. Except for negligible differences, all the results remain the same as before when $v(p)$ was uniform. If we pause to reflect on this fact, we find that it is not so surprising after all. As long as the number of experiments is small, the influence of the initial distribution predominates; however, as the number of experiments increases, this influence decreases more and more. Since Bayes's Theorem is, mathematically speaking, a proposition applying to an infinite number of experiments, we conclude that *the above-stated proposition of Bayes holds true independently of the given initial probabilities*, i.e., whatever the contents of the urn from which the stone was drawn.

Conditions like those given in our examples, such as an urn with a given distribution of objects, occur only very rarely. More commonly, we may pick a die about which nothing will be known except that it seems a suitable object with which to test the alternative 'six or nonsix'. The conclusions remain the same as before: If n experiments have shown n_1 successes, then, so long as n is small, *we cannot conclude anything from this* since for small n, the result of our inference depends mainly on the initial distribution, i.e., on the general make-up of the dice from which we have picked our particular die. If, however, n is a larger number, say 500, then we can draw conclusions *without any further knowledge* of the total body of available dice. And we can say that there is indeed a very high probability that the probability of success of the die we picked will lie close to the observed frequency of success. Once we have derived this fact, it appears clear and almost obvious; it is, however, quite remarkable to realize that it is a direct result of probability calculus and can be derived only on the basis of the frequency definition of probability.

A brief and useful formulation of Bayes's Theorem can be obtained by substituting for the term 'probability of success' the definition of this probability. We can then state: If a sufficiently long sequence of alternatives has shown a frequency of success a, we can expect with a probability very close to unity that the limiting value of this frequency will not be very different from a. This brings out clearly the close relation between the First and the Second Laws of Large Numbers.

In the fifth lecture, we shall discuss problems of mathematical statistics which are closely related to Bayes's Theorem. It is indeed the principal object of statistics to make inferences on the probability of events from their observed frequencies. As the initial probabilities of such events are generally unknown, that aspect of the inference problem which we have just discussed will prove to be essential. It explains why we can generally draw meaningful conclusions only from a large body of statistical observations and not from small groups of experiments.

It is not my intention to show how Bayes's Theorem is reduced, in the same way as Poisson's Theorem, to a purely arithmetical proposition if we adhere to the classical definition which regards probability as the ratio of the number of favourable events to the total number of equally possible events. This proposition leads to the prediction of empirical results only if we smuggle into it again an *ad hoc* hypothesis such as: 'Events for which the calculation gives a probability of nearly 1 can be expected to occur nearly always, i.e., in the great majority of trials'. This hypothesis, as we know, is equivalent to the frequency definition of probability in a restricted form. By introducing this additional hypothesis we change the meaning of 'probability' somewhere between the beginning and the end of the investigation.

From our point of view, a more important aspect of the question is the relation in the new frequency theory of probability between Bayes's Theorem and the Law of Large Numbers (Poisson's Theorem), and the relation of Bayes's proposition to the axiom of the existence of limiting values of relative frequencies. At first sight nothing seems more reasonable than to identify the proposition: 'If a relative frequency a is found in a sequence of length n (n being a large number), it will almost certainly remain nearly unchanged by an indefinite prolongation of the sequence', with the simple assertion: 'The relative frequency approaches a definite limiting value with increasing n.' The essential difference lies, however, in the words 'almost certainly', and these words can be replaced by 'nearly always'. If a stone has been drawn from the urn and thrown n times (n being a large number), and the result 6 has been obtained n_1 times, so that the relative frequency of this result is $a = n_1/n$, this experiment says nothing about the behaviour of the same stone in a prolonged sequence of throws. If we merely assume the existence of limiting values, and nothing concerning randomness, it is quite possible that for almost all stones which have given the same value of $n/n_1 = a$ in n throws the limiting value of the frequency will differ considerably from a, however large n may be.

Bayes's proposition means that in practice this is not the case, and that the limiting value is nearly always only slightly different from the observed relative frequency n_1/n. Bayes's Theorem thus contains a new statement, not identical with the premise used in its derivation (i.e., the first axiom) and obtainable only by a mathematical deduction, which uses also our second axiom, that of randomness. Because of the analogy of this theorem with the Law of Large Numbers, it is often called the 'Second Law of Large Numbers'.

These considerations have led us further than I intended into the field of abstract arguments, and I will conclude by restating the two Laws of Large Numbers and the First Axiom in the special form adapted to the problem of throwing dice.

THE THREE PROPOSITIONS

In n casts of a die, a certain result, say 6, was obtained n_1 times. The three propositions are as follows:

(1) *The First Axiom*, which is basic to our definition of probability, says that in an indefinitely long repetition of the same game the quotient n_1/n will approach a constant limiting value, this value being the probability of casting 6 with this particular die.

(2) *The First Law of Large Numbers*, which is also called the Bernoulli-Poisson Theorem, says that if the game of n casts is repeated again and again with the same die, and n is a sufficiently large number, *nearly all games* will yield nearly the same value of the ratio n_1/n.

(3) *The Second Law of Large Numbers*, or Bayes's Theorem, says that if, for a great number of different dice, each one has given n_1 results 6 in a game of n casts, where n is a sufficiently large number, *nearly all of these dice* must have almost the same limiting values of the relative frequencies of the result 6, namely, values only slightly different from the observed ratio n_1/n.

These formulations exactly delimit the three propositions; all that must be added is that the first of them is an axiom, that is to say an empirical statement which cannot be reduced to simpler components. The other two are derived mathematically from this axiom and the axiom of randomness. Propositions (2) and (3), the two Laws of Large Numbers, lose their relation to reality if we do not assume from the beginning the axiom (1) of limiting frequencies.

GENERALIZATION OF THE LAWS OF LARGE NUMBERS

The two theorems discussed, that of Bernoulli and Poisson and that bearing the name of Bayes, form the classical Laws of Large

Numbers. It may be worth while mentioning that we owe to Bayes only the statement of the problem and the principle of its solution. The theorem itself was first formulated by Laplace. In recent times the two laws have been considerably supplemented and extended. We are not going to discuss these contributions in detail here, because they do not affect the foundations of the theory of probability. On the other hand, we cannot pass over them altogether. Some of them are important for the practical application of the theory; others have caused a controversy which is interesting from our point of view. Although the position is absolutely clear, many opponents to the frequency theory still try to interpret every new theorem of this kind in such a way as to construct a contradiction to the frequency conception. This is especially true for the so-called Strong Law of Large Numbers.

Let us first consider the kinds of propositions which can reasonably be considered to belong to the class of Laws of Large Numbers. The limitations are, of course, more or less arbitrary. In accordance with a certain established usage, we suggest the following: First of all, the problem under consideration must contain a large number n of single observations; this is to say, it must represent a combination of n collectives. The aim of the calculation must be a probability P, determined by the n observations. In other words, the final collective must be produced by the combination of the n initial collectives, P being therefore a function of n. The last characteristic of the problem is the fact that P approaches 1 as n increases indefinitely. The solution can therefore always be formulated in a sentence beginning with 'It is almost certain that when n becomes very large . . .' In the theorem of Bernoulli and Poisson, it is 'almost certain' that the relative frequency of an event in a sequence of n single experiments is nearly equal to its probability. In the theorem of Bayes it is 'almost certain' that the unknown probability lies close to the frequency found in an empirical sequence. Of course these abbreviated formulations do not express the full content of the theorems; complete formulations have been given in the preceding paragraphs.

The word 'condensation' is a very adequate one for the description of the mathematical facts expressed in the different Laws of Large Numbers. Usually a probability distribution for a number of different attributes is more or less uniform. In the cases to which the Laws of Large Numbers apply, the total probability 1 is 'condensed' in one point or rather in its very close neighbourhood. The condensation becomes more and more pronounced with an increase in the value of the parameter n. Analogously to the terminology used in analysis, attempts have been made to introduce into the theory of probability the notion of convergence. In my opinion these attempts are not too fortunate. The expression 'convergence in the sense of probability' does not add much to the clarification of the facts. This terminology has been suggested by those mathematicians whose point of view I have characterized at the end of my preceding lecture, and who are inclined to make the theory of probability a part of the theory of sets or of the theory of functions.

THE STRONG LAW OF LARGE NUMBERS

Various mathematicians, especially Cantelli and Pólya,[8] have recently derived a proposition which goes somewhat further than the theorem of Bernoulli and Poisson. As in the derivation of Bernoulli's Theorem, we again consider a group of n repetitions of a simple alternative (e.g., 0 or 1), and treat this whole group·as an element in a collective composed of an infinite number of such groups. We have considered the number n_1 of 0's as the attribute of each group of n observations. If we introduce now the letter x to denote the relative frequency n_1/n of zeros in a group, the theorem of Bernoulli and Poisson says that it is almost certain that the frequency x is not very different from the probability p.

Let us now consider more closely the n single observations forming an element of our collective. Let m be a number smaller than n, so that $n - m = k$ is a positive number. Among the first $(m + 1)$ results in a group there will be a certain number of zeros, which may be anything between zero and $m + 1$. We denote the frequency of 0's in this part of the group by x_1. The corresponding frequency in the first $m + 2$ observations we denote by x_2, and so on, up to x_k. If, for instance, $m = 10$ and $n = 15$, we start by calculating the frequency of 0's within the first eleven observations. If six 0's have been recorded in the first eleven trials, then $x_1 = 6/11$. If the twelfth trial gives a 0, then $x_2 = 7/12$; in the opposite case $x_2 = 6/12$. The last frequency to be calculated is x_5, determined from the total number of 0's in the fifteen results. In fact, x_k is nothing else but the frequency denoted previously by x.

As attributes of an element in our collective, we now consider not simply the value of x or x_k, but the whole system of values x_1, x_2, . . ., x_k, which are all positive and less than 1. We shall now perform a mixing operation: Let p be the probability of a zero in the original simple alternative (e.g., $p = 1/2$ in the game with an ordinary coin), and ε a small positive number. Some of the k numbers x_1 to x_k may

127

belong to the interval from $p - \varepsilon$ to $p + \varepsilon$, others may lie outside it. If *at least one* of the k numbers x_1 to x_k falls outside this interval, we register this as an 'event'. If all the k numbers fall in the interval, we say that the event did not occur.

The collective derived in this way is again a simple and well-defined alternative. We are interested in the probability P of the occurrence of the above-defined event in a group of n single observations. We can describe P as the probability that the relative frequency of the result zero deviates by more than ε from the fixed value p at least once in the interval between the mth and the last (nth) single observations in a group of n experiments. This probability can be calculated by the repeated application of the rules described in the second lecture. The probability P depends on the four variables n, m, p, and ε. We are, however, not so much interested in the actual value of P, as in a certain remarkable property of this function. Calculation shows that P is *always smaller* than the reciprocal of the product of m and ε^2: i.e.,

$$P \text{ smaller than } \frac{1}{m\varepsilon^2}.$$

This relation is independent of the values of n and p.

Let us consider the meaning of this result more closely. However small the constant value adopted for ε (e.g., $\varepsilon = 0.001$, or $\varepsilon = 10^{-6}$), the expression $1/m\varepsilon^2$ decreases indefinitely with indefinite increase in m, and finally tends to zero. The number n, which is supposed to be larger than m, will of course also increase indefinitely during this process. When the probability of an event decreases to 0, the probability of its nonoccurrence increases to 1. The above relation can then be interpreted thus: It is almost certain that between the mth and the nth observations in a group of length n, the relative frequency of the event 'zero' will remain near the fixed value p and be within the interval from $p - \varepsilon$ to $p + \varepsilon$, provided that m and n are both sufficiently large numbers. The difference between this proposition and Bernoulli's Theorem is quite clear: formerly, we concluded only that the relative frequency x will almost certainly be nearly equal to p at the end of each group of n observations; now we add that, if m is sufficiently large, this frequency will remain nearly equal to p throughout the last part of the group, beginning from the mth observation.

The amazing and unexpected part of this result is the fact that the upper limit $1/m\varepsilon^2$ of the probability P is independent of n. This result has given rise to the incorrect and unclear explanations to which I have previously referred. Let us assume a constant value for m, say $m = 1000$, and consider steadily increasing values of n, say $n = 2000, 3000, 4000$, etc. Obviously, the probability P of the deviation in question increases with an increase in n, since the number $k = n - m$ of observations in which the deviation of the relative frequency from the fixed value p can occur becomes larger and larger. Calculation shows that despite this increase in the number of possibilities for the occurrence of the deviation, its probability will not increase above the fixed limit $1/m\varepsilon^2$. If ε is, for instance, 0.1 and m is 10,000, then $1/m\varepsilon^2$ is 0.01, and we can make the following statement: We may expect with a probability exceeding 99% that after 10,000 tosses of a coin the relative frequency of 'heads' will *always* be included within the interval from 0.4 to 0.6 (i.e., between $p - \varepsilon$ and $p + \varepsilon$), no matter how large the total number of tosses from which this frequency is calculated may be, whether it is a million, or ten millions, or a larger number still.

This formulation of the Strong Law of Large Numbers, and the way in which we derived it, shows clearly that both the problem and its solution fit into the general scheme of the frequency theory without difficulty. It is a problem of constructing a certain new collective by means of the usual operations. This is all I wanted to show; it is not my purpose to give a discussion of the incorrect expositions which the publication of this proposition provoked.

THE STATISTICAL FUNCTIONS

I wish to use the remainder of my time for the discussion of another generalization of the Law of Large Numbers,[9] which is of more practical interest and forms a suitable link with those problems of statistics with which we shall deal in the next lecture. We begin by substituting a general collective with many different attributes for the simple alternative (1 or 0) which we have been considering hitherto. As an example, we may use the game of roulette with its thirty-seven possible results, the numbers 0 to 36.

Let us consider, as an element in a new collective, a group of n, say 100, single games of roulette. The attribute of this new element is a certain arrangement of 100 numbers from 0 to 36. We may, however, not be interested in the 100 individual results, but only in the relative frequency of the 37 attributes, i.e., in indicating the number of times when 0, 1, 2, . . . and, finally, 36 appear in the group in question. We call this the statistical description of the group, and it is obvious that the sum of these 37 entries will be $n = 100$. If

129

we divide each of these entries by n, we shall obtain a sequence of proper fractions with sum 1. These fractions $x_0, x_1, x_2, x_3, \ldots, x_{36}$ are the relative frequencies of the occurrence of the different results, 0 to 36, in the group under consideration. These quantities $x_0, x_1, x_2, x_3, \ldots$, with the sum 1, describe what we call the *frequency distribution* of the various results in the group of 100 experiments.

In the sense of the definitions given in the second lecture, the transition from the original complete description of the group to the new abbreviated one is equivalent to a 'mixing' operation. A given frequency distribution can be due to a great number of different arrangements. In the simple case of only two possible results, 0 and 1, and a group of, say, 10 casts, the distribution of 0.30 zeros and 0.70 ones can be produced by 120 different arrangements, since three 0's and seven 1's can be arranged in 120 different ways. If the probability of each possible arrangement is known, the probability of any frequency distribution can be calculated according to the law of mixing, by means of summations.

It must be pointed out that the same frequency distribution can correspond to different lengths of the group. For instance, the distribution of three 0's, two 1's, and five 2's in a group of ten observations is equal to that of fifteen zeros, ten 1's, and twenty-five 2's in a group with $n = 50$. To say that we know the distribution of some attributes in a certain experimental material does not necessarily involve the knowledge of the number n of the experiments made.

The subject of our interest is, however, often not the frequency distribution as such, but some quantity derived from it. It is this quantity which is considered as the attribute in which we are interested. This amounts to the performance of another mixing operation, in which all distributions leading to the same value of the attribute are mixed. A quantity of this kind, derived from n single results, but depending only on their frequency distribution (and not on their arrangement or on the total number n), is called a *statistical function*.

The simplest example of a statistical function is the average of the observations. If fifteen 0's, ten 1's, and twenty-five 2's have been counted in a group of fifty results, the average is calculated by dividing by fifty the following sum:

$$(15 \times 0) + (10 \times 1) + (25 \times 2) = 10 + 50 = 60;$$

this gives us the result $60/50 = 1.20$. Another method of calculation, which leads to the same result, is to multiply each of the three results (0, 1, 2) by its frequency in the distribution; in our case, the three frequencies are 0.30, 0.20, and 0.50, and the addition of the three products gives

$$(0.30 \times 0) + (0.20 \times 1) + (0.50 \times 2) = 0.20 + 1.00 = 1.20.$$

We see that the average depends in fact only on the numbers x_0, x_1, and x_2, and not on the arrangement of the fifty results. Neither is it affected by doubling the total number of observations, provided that this involves doubling the number of results of every kind.

The average, or mean, of a number of results is thus a statistical function. Various other statistical functions will be described in the next lecture. We are now going to say a few words on the Laws of Large Numbers in their relation to statistical functions.

THE FIRST LAW OF LARGE NUMBERS FOR STATISTICAL FUNCTIONS

The average of a sequence of n measurements whose single results are either 0's or 1's is obviously equal to the ratio of the number of 1's to n. The Bernoulli Theorem may therefore be stated in this way: If n observations of an alternative (0 or 1) are grouped together to form an element of a collective, it is almost certain that the average of the n observations will be nearly equal to a certain known number p, if n is sufficiently large.

Poisson's generalization of Bernoulli's proposition can now be formulated as follows: The n observations need not be derived from a single alternative; it is also admissible to calculate the average in the case of n different alternatives, that is, to divide by n the number of 'positive' results obtained in such a sequence of experiments. The 'condensation of probability' demonstrated by Bernoulli for the case of n identical alternatives occurs in this case as well.

The next step in the generalization of the theorem is mainly due to Tschebyscheff. Tschebyscheff's proposition says that the results need not be derived from simple alternatives; they can also be taken from collectives with more than two different attributes. We can, for instance, consider n games played with a certain roulette wheel, or n games played with different wheels, and calculate the average of the n numbers representing the results of all games. If n is sufficiently large, it is almost certain that this average will be nearly equal to a certain known number, which is determined by the probability distributions of the n initial collectives.

Certain recent investigations enable us to generalize the proposition still further, and in this new form it becomes a very important theorem of practical statistics. The phenomenon of 'condensation', first described by Bernoulli, holds not only for the average of n results, but essentially for all statistical functions of n observations, if n is a sufficiently large number. In other words, if we observe n collectives (which may be all equal or different), and if we calculate the value of some function of the n results which depends on their frequency distribution (but not on the order of the observations nor on their total number), then, provided n is sufficiently large, it is almost certain that the value so obtained will differ but little from a certain known number, calculated in advance from the probability distributions of the n collectives. The expressions 'almost certain', etc., are, of course, abbreviations whose meaning must be interpreted, according to our definition of probability, in the following way. If a group of n experiments is repeated a very large number of times, and if ε is an arbitrary small number, the value calculated each time from the n observations will in the overwhelming majority of all groups differ by less than ε from the 'theoretical' known value. The larger n is, the greater is the majority of cases in which this prediction turns out to be true for a given constant value of ε.

THE SECOND LAW OF LARGE NUMBERS FOR STATISTICAL FUNCTIONS

Bayes's Theorem, discussed in one of the preceding sections, can also be generalized in such a way as to apply to statistical functions. Let us imagine now that n observations have been made on *one* collective whose distribution is unknown, e.g., n throws with a stone of apparently cubic form and selected at random from a heap of such stones. From n observations made with this stone (each characterized by one of the numbers 1 to 6), we deduce a certain number L, which depends neither on the order of results nor on the value of n, but only on the frequency distribution of the results; L is, therefore, a statistical function. According to the considerations of the preceding sections, we must assume the existence of a certain "theoretical value of L", denoted by L_0, which is determined by the probability distribution in the original collective, in our case, by the six probabilities of the six sides of the stone; L_0 is unknown, since we have assumed that the stone which served for the experiments has not been investigated before. All we know about this stone is the value of L calculated from the observations.

The Second Law of Large Numbers, generalized for statistical functions, asserts that, with a sufficiently large value of n, the unknown theoretical value L_0 will 'almost certainly' lie quite near to the observed experimental value L. The original Bayes Theorem was equivalent to the above statement for the special case of the average of the results of a simple alternative; the 'theoretical' value was in this case the fundamental probability p of the event under consideration. We can now formulate the general proposition: Assuming that n observations have been made on an unknown collective and have produced the value L of a certain statistical function, if n is a sufficiently large number it is almost certain that the unknown theoretical value of this function, L_0, characteristic of the distribution in the collective under investigation, lies very near to the observed value L. The way of interpreting the expressions 'almost certain', etc., by means of the frequency definition of probability has been indicated above.

The proposition which we have now formulated thus allows an inference into the nature of an unknown collective based on the results of a sufficiently long sequence of experiments. It is therefore of an even greater practical importance than the First Law of Large Numbers. It is perhaps the most important theorem of theoretical statistics. In the next lecture we shall consider as an example of an important statistical function the so-called Lexis's Ratio. If a sufficiently long sequence of observations has given for this ratio a certain value, say $L = 1.1$, the generalized Second Law of Large Numbers allows us to assume that the 'theoretical' value of L (i.e., the value characteristic of the material of our investigation) is nearly equal to 1.1. I do not intend to deal here with the mathematical characterization of those statistical functions to which the two laws apply. Instead, I will close this lecture with the following remarks.

CLOSING REMARKS

In this lecture I have tried to elucidate a number of questions connected with the popular ideas of the Laws of Large Numbers, as far as this is possible with a minimum use of mathematics. I had in mind a double purpose: In the first place, I wanted to acquaint you with the content of these laws, which are indispensable to anyone who has to deal with statistical problems in one form or another, or with other applications of the theory of probability. Secondly, it was especially important for me to investigate the part played by these laws in the calculus of probability based on the frequency definition. As I said at the beginning of this lecture, and also indicated in the previous one, many objections raised against my probability theory have been directed against its supposed contradictions to the Laws of Large Numbers. I hope that I have succeeded in making the following two points sufficiently clear.

(1) Starting with the notion of a collective and the definition of probability as a limiting value of relative frequency, all the Laws of Large Numbers have a clear and unambiguous meaning free from contradictions. Each of them amounts to a definite prediction concerning the outcome in a very long sequence of experiments, each of which consists of a great number n of single observations.

(2) If we base the concept of probability, not on the notion of relative frequency, but on the definition used in the classical probability theory, none of the Laws of Large Numbers is capable of a prediction concerning the outcome of sequences of observations. When such conclusions are nevertheless drawn, this is possible only if, at the end of the calculations, the meaning of the word 'probability' is silently changed from that adopted at the start to a definition based on the concept of frequency. Naturally, such a procedure may lead to obscurities and contradictions.

Before concluding, I must add another warning. It is impossible to give absolutely correct formulations of the propositions we have discussed if the use of formulas and mathematical concepts, except those of the most elementary nature, is avoided. I hope that I have succeeded in stating all the essentials correctly. From the mathematical point of view, the formulations which I have given are still incomplete, lacking various restrictions, such as those concerning the sets of admissible functions, as well as further formal mathematical conditions. Those who are interested and possess the necessary mathematical knowledge can find this information in the mathematical literature of the subject.

Applications in Statistics and the Theory of Errors

THIS lecture and the next, which concludes this series, will be devoted to a consideration of the two most important applications of the theory of probability. We shall no longer concentrate on games of chance. In the present lecture we will deal with various series of events, which occur in everyday life, and whose investigation is commonly called 'statistics'.

WHAT IS STATISTICS?

The word statistics has been interpreted as the 'investigation of large numbers', or 'theory of frequencies'. This is not the literal meaning of the word, but an attempt to make clear the sense which it has acquired in modern language. Long sequences of similar phenomena which can be characterized by numbers form the subject matter of statistics; examples are: *population statistics* (e.g., birth rates and death rates); *statistics of social phenomena* (e.g., marriages, suicides, incomes); *statistics of biological phenomena* (e.g., heredity, sizes of different organs); *medical statistics* (e.g., action of drugs, cures); *technological and industrial statistics* (e.g., mass production, mass consumption, most problems grouped today under the heading of operational research); *economic statistics* (e.g., prices, demand).

In these and similar fields, the usual procedure is to collect empirical material, draw general conclusions and use these to form further conclusions which may be applied in practice. Which part of this investigation should be called statistics is more or less arbitrary. We are not going to intervene in the struggle between different schools of thought, the 'general' and the 'mathematical', the 'formal' and the 'realistic'. All that is necessary for us is to delimit the field which we

135

propose to consider in these lectures. We leave aside all questions of planning and carrying out statistical investigations, such as the organization of a census, as well as measures which may be taken as the result of the outcome of such investigations. It is, of course, impossible to limit any field of research with complete rigidity. The doctor whose patient has broken a leg does not ask for details of his education or his future plans. He cannot, however, be completely disinterested in anything that is important for the physical or mental state of his patient. On the other hand, however, the way to any progress and increased effectiveness consists in properly fixing the object of one's attention and then proceeding to study it along varying lines of approach.

The scope of our interest in the field of mass phenomena and repetitive events is indicated by the theory presented in the foregoing lectures. It is true that the starting point of a theory is always the wish to classify and explain a certain group of empirical facts. However, once the theory has been developed, we usually find that it explains only one of the many aspects of reality and we must be satisfied with an understanding of this one side.

The starting point of all our statistical considerations is the concept of the collective, which is the abstract and idealized equivalent of the empirical notion of a mass phenomenon. It is impossible to sum up in two sentences the subject of a discussion which is to occupy fifty pages, but I will attempt to point out our particular approach to statistical problems: We shall consider different subjects of statistical interest and investigate *whether and to what extent they can be interpreted as collectives or reduced to collectives*. In cases when a connexion with a collective is found, the empirical data can be dealt with by means of our probability theory.

GAMES OF CHANCE AND GAMES OF SKILL

Before entering upon our subject, we must return for a few minutes to the problem of games of chance, and consider an aspect of this problem which can serve as an introduction to the questions of practical statistics. The public organization of games of chance is prohibited by law in most countries, or at least is subject to certain restrictions. To apply these laws, a criterion is required to distinguish games of chance from games of skill. In some of the older regulations games of chance were defined as 'games of aleatoric character': this is an 'explanation by substitution' of the worst kind, since it 'explains' a well-known word by one which is generally unknown but has a scientific appearance. Most modern legislation seeks no definitions of games of chance, assuming that the necessary distinction can be made in each particular case by the application of well-known general principles.

A game which is often played at fairs has repeatedly occupied the attention of the law courts. It consists in catching in a cup held by the player a ball rolling down a sloping board studded with pins. No agreement could be reached for a long time on the nature of this game, as to whether it is a game of chance or a game of skill. In many other cases the decision is much more difficult. Roulette is a game of chance, chess is a game of skill. These extreme cases raise no question. In all card games, even in those which nobody would consider games of chance, chance intervenes, if only by the random distribution of cards to the players. All kinds of intermediate cases between games of pure chance and those of pure skill are obviously possible, resulting from the combination of certain elements of chance with those of skill and wit required by the rules of the game.

A reasonable definition can be derived directly from our concept of a collective. A game of pure chance is a game in which the players' chances of winning, that is, the relative frequencies of wins in an infinitely long sequence of games, do not depend on their personal qualities. In each practical case (excepting games of a purely mechanical character which work automatically without participation of the players), the final decision can be reached only by means of a sufficiently long statistical experiment, i.e., a prolonged sequence of games. No *a priori* decision, i.e., a logical deduction based on the rules of the game, is possible. If sometimes the necessity for an experimental test is overlooked, it is only because this test has been carried out so often that its results are well known to everyone. The game of roulette is an example of this kind. In accordance with the position reached in the preceding lectures, the randomness of the results of the games of dice and roulette is nothing but a postulate borne out by experimental results. In this way the whole field of games of chance finds uniform treatment.

For the purpose of the legislator, however, it would be better not to confine our definition to games of pure chance. An appropriate formulation of the law might be as follows: All games are prohibited in which the distribution of wins is found in the long run to depend only very slightly on the skill of the individual players or to be entirely independent of it.

In the case of the above-mentioned game at the fair, the path taken by the falling ball is directed by chance; but the success in

137

catching it depends essentially on the speed and correct judgment with which the cup is placed in the right position by the player, in fact, on his skill. To make this game one purely of chance, one might require that the cup must be brought in position before the ball starts rolling, different prizes being assigned to different positions.

MARBE'S 'UNIFORMITY IN THE WORLD'

I now leave the games of chance and turn to a borderline question between biological and social statistics, which is closely related to problems of games of chance. This subject has been discussed in a comprehensive work entitled *Uniformity in the World* (and in certain other works), the German philosopher Karl Marbe[1] developed a theory which can best be illustrated by an example.

A husband, whose wife is imminently expecting a child and who is anxious for male issue, goes to the registrar of births and looks up how many boys and girls have been born in the last few days. If the number of girls has been comparatively large, he considers that he has a good chance of becoming the father of a son. The general opinion is that this line of thought is foolish. Marbe, however, says: 'Our statistical investigations show (and this fact has so far been overlooked) that the expectation of this father is not absolutely unfounded'. Marbe supports this conclusion by statistical data concerning about 200,000 birth registrations in four towns of Bavaria, which show only one 'run' of seventeen consecutive births of children of the same sex, and not a single such run of greater length. The study of long runs is also the basis of most attempts to arrive at a gambling system. In his book Marbe actually gives suggestions on how to improve the chances of winning at Monte Carlo by an appropriate selection of games.

Marbe's Problem has been the subject of many discussions, and Marbe himself has added two supplements to his original two-volume work.[2] It interests us in so far as it is related to the foundations of the theory of probability. Marbe's idea is obviously that once seventeen male children have been born in succession in a town, the probability of the next child being a female becomes a practical certainty. If we should consider the successive birth registrations as a collective with the alternative properties 'F' (female) or 'M' (male) (similar to the results in a game of 'heads or tails'), the principle of randomness would lead us to expect the distribution of M's and F's in the partial sequence formed by all registrations following seventeen consecutive M's to be exactly the same as in the collective as a whole. This selection is in fact nothing but an ordinary 'place selection', which we assumed did not affect the relative frequencies of different attributes in a collective. The question arising from Marbe's suggested 'statistical stabilization' is thus: Does the sequence of birth registrations, with the sex as the attribute, possess the properties of a collective? It is obvious that the answer can be found only by empirical methods. All attempts to disprove Marbe's theory of statistical stabilization by a priori considerations or logical arguments are useless. The problem whether a given group of statistical results should or should not be considered as a collective lies at the centre of most difficulties of practical statistics.

ANSWER TO MARBE'S PROBLEM

It must be stated first of all that, despite the enormous quantity of statistical data compiled by Marbe, he has not brought forward any stringent proof of his assertion. A run of length 17 has occurred only once in his material; that means that he has only one example by which to judge what happens after a run of this length. Thus, the partial sequence of registrations following runs of 17 male births consists so far of one element only, whereas conclusions relative to the statistical distribution in a sequence can be drawn only after the sequence itself has reached a considerable length. On the other hand, neither does this result say anything *against* Marbe's hypothesis. Other facts are therefore required for the confirmation or refutation of Marbe's theory.[3]

Our theory enables us to devise various tests by which mathematical conclusions can be compared with the results of observations. The possibility of such indirect arguments is one of the essential advantages of any theory. By means of the fundamental operations of the theory of probability, we can derive from the given collective with the attributes M and F and approximate distribution 0.5/0.5, a new collective whose elements are sequences of n observations each, and whose distribution determines the probability of encountering exactly x runs of length m in a finite sequence of length n. By introducing the special values $m = 17$ and $x = 1$ into the formula obtained in this way, we obtain the probability of one single run of length 17 being found in a certain number of experimental data. Using $m = 18$ and $x = 0$, we can calculate the probability of a complete absence of runs of length 18 in the same material. These results are interesting in connexion with Marbe's views, since the occurrence of just one run of length 17 and the absence of any longer runs were the main arguments for advancing his hypothesis. If we could show that these results are in accordance with the predictions of probability calculus, this would be a strong argument for the randomness of the distribution of male and female births, i.e., for the hypothesis that the birth register shows the characteristic properties of a collective.

The numerical calculation for $n = 49,152$ (this being the length of one of the four sequences investigated by Marbe) gives, for a run of length 17, the following probabilities: 0.16 for one occurrence; 0.02 for two or more occurrences; 0.82 for no occurrence. In one of the four sequences of Marbe, the iteration 17 occurred once; in the other three it did not occur at all. Theory and experiment are thus, to say the least, not in contradiction to each other. For the run of length $m = 18$, the calculated probability of occurrence (single or repeated) is 0.09; the probability that such a run will not occur at all is thus 0.91. The fact that no run of length 18 was observed in four sequences is in accordance with this theoretical prediction. We have no space here to enter into more details concerning the analysis of the experimental data and their comparison with the predictions of the theory of probability. Suffice it to say that the result of this comparison is a complete confirmation of all the conclusions drawn from the assumption that the birth register is a collective. As examples, I might quote the expected numbers of the runs of lengths 2, 6, and 10, respectively. They are 6138, 385, and 24.3, whereas the corresponding average numbers in the four sequences analysed by Marbe were 6071, 383, and 24.3. I have discussed the details in one of my earlier papers. The more recent of Marbe's assertions, although more subtle, point in the same direction as the original ones and can be refuted by calculations analogous to those just reported.

Everybody interested in the foundations of probability must be grateful to Marbe for collecting and organizing such a large amount of statistical material. It provided the theory of probability with the first example of a large collective not taken from the field of games of chance. The conclusions which we draw from this material are, however, diametrically opposite to those drawn by Marbe himself. The very good agreement of the results of the probability calculations with the statistical data shows us that the register of births approximates to a high degree the properties of a collective, in particular that of randomness.

Returning now to the fathers who may expect the birth of a son because they have found that 17 girls have just been born in succession (few fathers will ever be in this position), we conclude that about

50% of them will be disappointed and will return to the registrar to add another F to the run of female births, and about 50% will break the run by registering a male birth.

THEORY OF ACCUMULATION AND THE LAW OF SERIES

A peculiar counterpart of Marbe's Theory of Statistical Stabilization is the 'Theory of Accumulation' propounded by another philosopher, Sterzinger.[4] A biologist, Paul Kammerer,[5] adopted and used it as a scientific basis for a 'Law of Series', which is a conception widely accepted by the public.

A proverb says 'troubles seldom come singly'; and Kammerer assumed that the same must be true for all kinds of other events as well.

More precisely, the Law of Series amounts to the assertion that short runs occur unexpectedly frequently, supposedly more frequently than they should according to the theory of probability. For example, Sterzinger observed the intervals at which buyers entered a shop. On the average, 30 buyers arrived each hour; it was, however, quite rare that *one* buyer should enter the shop in an interval of two minutes. Most 2-minute intervals were either 'empty', or 2 or more persons entered the shop in this time. Sterzinger drew far-reaching conclusions from observations of this kind; he saw in them a disproof of the applicability of the theory of probability to actual phenomena. In considering this question our point of view follows directly from our general conception of probability. Mass phenomena to which the theory of probability does not apply are, of course, of common occurrence. In other words, not all repetitive events are collectives in the theoretical sense of the word. An analysis of the experimental material is necessary to show whether a given mass phenomenon fulfils the conditions of a collective or not; but superficial, instinctive estimates are of no value. In order to point out a valid deviation of the experimental result from the theory, it is above all necessary to derive from the theory mathematical conclusions which will permit a comparison with experience.

An example of such a conclusion is the following: If a sequence of identical events is distributed at random over a certain period of time and one event occurs on the average every a minutes, then the probability of an arbitrary time interval of the length of a minutes witnessing one (and only one) event is 0.37 while 0.63 of all intervals will either be 'empty' or contain two or more events. Therefore Sterzinger's experiments would be a valid argument in favour of the accumulation theory only if he should have found more than 2/3 of the 2-minute intervals empty or having 2 or more buyers enter the shop under observation. This could certainly happen and if it were observed it would indicate that this sequence of observations does not satisfy the conditions of a collective. Unfortunately, Sterzinger's actual observations are entirely insufficient for deciding this question. His own opinion was obviously that the mere fact of the repeated occurrence of 2-minute intervals with less or more than 1 buyer is already a decisive argument against the validity of the theory of probability.

Here is a further example of an erroneous application of the Law of Series: Statistics showed the mean weekly number of suicides in normal times in Germany to be 1 per 250,000 inhabitants. How often are we to expect the newspaper in a town of 250,000 inhabitants to report a 'series of suicides', namely, three or more in one week? Calculation shows that the probability of more than 2 events of this kind occurring in one week is 0.08. If we assume that the register of suicides has the properties of a collective, we can expect the number of suicides per week to be 3 or more in about 4 weeks of each year (52 × 0.08 = 4.16). Only if a much greater frequency of 'suicide series' were observed would there be a reason for speaking of an 'epidemic of suicides' as is usually done in such cases. An actual occurrence of an 'epidemic' of this kind would point to some internal relations between the single cases. Kammerer's voluminous work does not contain a single example in which the statistical material is sufficient to decide the question whether the corresponding sequence of events has the characteristic properties of a collective or not.

LINKED EVENTS

Other statistical investigations have produced much evidence of repetitive events which are in fact not simply elements in a collective, but may be considered as obeying a certain law of series. We prefer, however, not to use this term, which historically has always implied a contradiction with the theory of probability. No contradiction exists, if the events are correctly interpreted. A typical example of a sequence of events of this kind, which we call 'linked', are deaths from a contagious disease, such as smallpox.[6] It is a priori clear that the distribution of these events must be fundamentally different from for instance, that of deaths by suicide. The mathematician G. Pólya has developed an elegant method for treating such sequences of linked events according to the principles of the theory of probability. The distribution of deaths due to some noncontagious disease can be illustrated in this way: An urn contains a number of white and black balls. Every month a ball is drawn from the urn by each inhabitant of the town. A white ball means life, a black one, death from this disease. The proportion of black balls in the urn is determined by the frequency of the deaths due to the cause under consideration. This "scheme of urns" means only that the probability of the occurrence of a certain number x of deaths can be determined in the same way as the probability of drawing x black balls from the urn. After the draw, the ball must be returned to the urn before the next draw takes place.

This picture ceases to be correct if the cause of death is in any way contagious. The occurrence of a single death increases in this case the chances for the repetition of this event. According to Pólya, this situation may be illustrated in the following way: Each time a black ball is drawn from the urn, not only is the same ball replaced, but a certain number of new black balls is added. If we adopt this representation we are no longer dealing with one given initial collective but with several collectives. However, the distribution in each of them is known in principle, since it depends only on the original proportion of black balls and the number of black balls added each time. By applying certain fundamental operations repeatedly, we can derive a final collective, whose elements are sequences of n draws, n being the number of inhabitants. Pólya has shown that in this way a comparatively simple formula is obtained which permits the calculation of the probabilities of different numbers of black balls in n draws, or of 0, 1, 2, 3, . . . deaths per month being caused by the disease under consideration.

The following figures from the Swiss statistics of deaths from smallpox from 1877 to 1900 show the agreement of theory and observation. The average number of deaths per month was 5.5; no deaths were recorded in 100 months, 1 death in each of 39 months, 2 deaths in each of 28 months, 3 in each of 26 months, 4 in 13, . . ., 15 in 3 months, and so on. Pólya's theory of linked probabilities leads, under appropriate assumptions, to the numbers: 100.4 months in which no deaths would be expected; 36.3, 23.5, 17.5, 13.8 months with 1, 2, 3, 4 deaths, respectively; and 3.0 months with 15 deaths each. Without linkage, that is, according to the ordinary theory of independent events corresponding to draws from an urn where each ball is replaced immediately, the numbers would be: 1.2 for no death; 6.5, 17.8, 32.6, 44.9 for 1 to 4 deaths, respectively; and 0.1 for 15 deaths. Comparison of the two results illustrates strikingly the superiority of the correct theory over the primitive one. It is difficult to realize the full influence of the contagion on the distribution of deaths without the help of a quantitative theory. For this influence does not merely consist, as one might expect, in an enhanced frequency of long runs of deaths, but it changes the whole distribution of runs, from the shortest to the longest.

We shall have to return to the effect of linkage (or 'after-effects') in the next lecture, when we shall speak of certain phenomena in physics in which these effects occur in a still more general way.

THE GENERAL PURPOSE OF STATISTICS

We have reviewed a number of special cases which enable us to understand a little better the general problem of the treatment of statistical results by means of the theory of probability. I have already mentioned that Marbe's Problem brings us close to the central problem of statistics, that of finding out whether a certain group of statistical data can be considered as a collective or not. The general problem can be formulated thus: A certain set of statistical data has been collected, such as data concerning the number of marriages in different parts of the country in a sequence of years. The question then arises whether this set of figures has the properties of a collective, or, more exactly, the properties of a finite sequence forming part of an infinite sequence with the properties of a collective. If this is not the case, can the group be reduced in one way or another to one or more collectives? An example of the latter type was the statistics of deaths from smallpox which we discussed in the preceding section. The observed sequence of figures in this case did not possess the quality of randomness required of a collective; it was, however, possible to reduce these figures to a certain system of collectives. Such a reduction enables us to make certain predictions concerning future events. Later, I shall show by means of other examples what can actually be achieved in this direction by the theory of probability. First, however, I should like to defend the above statement of the purpose of theoretical statistics against certain objections which I feel can be raised against it.

The concept of a collective was first derived from the analysis of statistical data. For example, in the first lecture, when I defined the probability of death, I used the statistical data concerning the number of deaths of insured persons. It might be said that it is surely not logical to say now that the fundamental problem of statistics is to investigate whether such systems of empirical data are collectives or not. This is, however, only a superficial objection, and everybody acquainted with the methods of exact science can easily resolve the apparent contradiction.

Newton's theory of gravitation led to the well-known conclusion that the orbits of the planets are ellipses with the sun at the focus. Kepler's determination of the elliptical character of the orbits of the planets was one of the starting points of Newton's theory. We know, however, that not a single one of the nine large and several thousand small planets revolves in an exactly elliptical orbit. One of the most important problems of mathematical astronomy is to explain all observed deviations from the elliptical paths by means of forces acting in accordance with Newton's law. In the same way, we are convinced that Maxwell's equations of electrodynamics correctly describe all electrical and magnetic phenomena in the world (with the exception of those involving very rapidly moving media). The explanation, by means of this general theory, of the details of the processes occurring, for instance, in different electrical machines nevertheless remains an important, practically inexhaustible, part of theoretical electrodynamics. To state that these 'mental experiments' underlying Maxwell's concepts and equations do not directly correspond to some reality does not amount to an objection to his theory. We can only expect to find in nature a certain approximation to the idealized conditions assumed by Maxwell in his deductions.

The objections to the theory of probability which point out that data found by statistical investigations never correspond exactly and often not directly to the idealized sequences called collectives must be met in the same way. An exact identity between theoretical premises and real conditions is not required, but only a similarity which makes a successful application of the theory to empirical data possible. The question that interests us is, therefore, what can be achieved by practical application of the theory of probability founded on the abstract concept of a collective?

LEXIS'S THEORY OF DISPERSION

The German economist, W. Lexis,[7] has enriched statistics with an idea which leads to a very convenient method of comparing an observed statistical sequence of figures with a collective. I will explain Lexis's Theory by means of a small statistical investigation which every one of you can easily repeat.

The subject of the investigation is the frequency of occurrence of the letter '*a*' in a Latin text, the first chapter of Cæsar's *De Bello Gallico*. In considering the first 2000 letters, divided into 20 groups

145

of 100 letters each, we find 2 groups with 5 a's in each, 3 groups with 6, 3 with 7, 3 with 8, 2 with 9, 1 with 10, 2 with 11, 2 with 12, 1 with 13, and, finally, 1 with 14. Altogether the number of a's is 174, and the mean frequency of this letter in a group is $174/20 = 8.7$. The question which we wish to answer is: How does this statistical result compare with the result which we would obtain by placing in an urn, in an appropriate proportion, a large number of tickets, with one of the 25 letters of the Latin alphabet printed on each, and making 100 draws of 20 tickets each? The appropriate proportion of a tickets would be 8.7%, since 8.7 was the mean frequency of this letter in our experiment.

To solve this problem, we could proceed by first calculating the distribution in a collective whose elements are series of 100 draws from the urn, and whose attributes are the different possible numbers of a's in a draw. We could then compare this computed probability for 0, 1, 2, 3, . . . a's with the empirical relative frequencies we have quoted. In this way we find, for instance, for the probability of 6 a's the value 0.10, for that of 8 a's, 0.14; we should therefore expect theoretically $20 \times 0.10 = 2.0$ groups of the first kind and $20 \times 0.14 = 2.8$ groups of the second kind, in a total of 20 groups. Both kinds of groups occurred actually 3 times in the empirical table. Is this a confirmation of the theory or not? It is difficult to answer this question. First of all, one should not be satisfied with comparing the results concerning these two kinds of groups; the calculations should be extended to all possible frequencies, i.e., to groups containing from 0 to 100 a's. To judge correctly the comprehensive table of results obtained in this way, one obviously needs some unambiguous measure to express quantitatively the degree of agreement or disagreement between theory and experiment. We need some kind of average of the deviations of the experimental values from the theoretical ones. Lexis's Theory attempts to define such a measure. It enables us to express the relation of theory and experiment by a single number.

THE MEAN AND THE DISPERSION

To explain Lexis's Theory we must first describe the meaning of the notion 'dispersion', applied to a sequence of numbers. Dispersion is a statistical function of the kind discussed at the end of the preceding lecture, and one of the most important among them.

I assume that all of you know that by adding the single results obtained in measuring repeatedly a physical magnitude and dividing the sum by the total number of measurements we obtain the so-called 'average' of the magnitude in question. We have already said on a previous occasion that the average is the simplest example of a statistical function. In the above example, the 20 results ranged from 5 to 14, and the average was 8.7. We can now calculate the difference between the average and the single values used in its calculation. We obtain in this way a new sequence of 20 numbers, partly positive and partly negative. The value 7 occurred three times in the original sequence; the sequence of differences therefore contains the number − 1.7 three times. The value 11 occurred twice; the sequence of differences contains two numbers 2.3, and so on.

The method by which the average, or mean, is calculated assures that the sum of all the positive and negative differences must be zero. The list of differences occurring in the above-discussed example is as follows:

No. of occurrences	Differences
2	− 3.7
3	− 2.7, − 1.7, and − 0.7
2	0.3
1	1.3
2	2.3 and 3.3
1	4.3 and 5.3

Thus the sum of the negative differences is $2 \times 3.7 + 3 \times (2.7 + 1.7 + 0.7) = 7.4 + 15.3 = 22.7$; the sum of the positive differences is the same. It is therefore obvious that the arithmetical mean of the deviations from the average cannot be used as a measure of variability of the single observations. The simplest way to arrive at an appropriate measure of the variability of the results is to square the single deviations and to sum these squares, which are all positive numbers. We define the 'dispersion' or the 'standard deviation squared' of a sequence of numbers as the arithmetical mean of the squares of all deviations from the average. Clearly, this quantity is a good measure of what we should intuitively call the variability of the given sequence of numbers. The smaller this variability, i.e., the closer the single numbers to their mean, the smaller is the dispersion. When all the given numbers are equal, they must be equal to their mean, and their dispersion is zero; and obviously, this is the only case when this happens. In our example we have to add two squares of − 3.7 to three squares of − 2.7, − 1.7, and − 0.7 respectively; this gives $2 \times 13.69 + 3 \times (7.29 + 2.89 + 0.49) = 27.38 + 32.01 = 59.39$. We have further to add twice the squares of 0.3, 2.3, and

3.3, as well as the squares of 1.3, 4.3, and 5.3 (together 48.27). The sum total of all the squares is 140.20, and $140.20/20 = 7.01$. This is the dispersion of the results under consideration.

It can easily be shown that the dispersion defined above possesses the characteristic properties of a statistical function. In fact, the order of the single results is obviously irrelevant for the value of the dispersion; if the total number of observations is doubled and each observed value occurs twice as often as before, the denominator and the numerator of the fraction used for the calculation of the dispersion are both doubled and therefore the value of the dispersion itself remains unchanged.

COMPARISON BETWEEN THE OBSERVED AND THE EXPECTED VARIANCE

Lexis now compares the dispersion of the observed results, i.e., the empirical value of the variance, obtained in the way we have described, with a theoretical value calculated for the corresponding collective by means of the theory of probability. Let us draw 100 tickets from an urn containing 87 a tickets in 1000. If similar draws of 100 tickets are repeated a great number of times, each group of draws can be considered as an element in a collective, with the number of a's as the attribute. By applying the operation of combination to 20 such elements we form a new collective whose elements consist now of 20 groups of 100 draws each. Any magnitude which in a given way depends on the 2000 results arranged in 20 groups of 100 draws can be considered as the attribute in the newly formed collective; e.g. the dispersion of the 20 values of the frequencies of the letter a is such an attribute. By the repeated application of the operations which we have discussed in the preceding lectures, it is possible to calculate the distribution in the new collective and to find the probabilities of the possible values of this dispersion. For instance, it would be possible to calculate the probability of the value 7.01 which we have actually found for the dispersion.

There is, however, another method of comparison which is more appropriate; this we shall now describe. We consider a collective with attributes represented by numbers. By multiplying the numerical value of each attribute by its probability and adding the products, we obtain the 'expected value' of the attribute, which is also sometimes called the 'mathematical expectation'.

A gambler who has 20% probability of winning 10 shillings in a game of chance, 30% probability of winning 20 shillings, and 50%

of winning nothing can say that his mathematical expectation is to win $0.20 \times 10 + 0.30 \times 20 = 8$ shillings. The meaning of this expected value is clear: it is the average win in an indefinitely long sequence of games.

There is an algebraic method for calculating the 'expected value' of an attribute in a collective of the kind under discussion which is formed by successive combinations of a number of initial collectives. Let n be the number of groups (in our case, $n = 20$), z the number of games in a group (in our case, $z = 100$) and p the probability of the result in question in a single group (in our case $87/1000 = 0.087$). Under these conditions, the expected value of the dispersion is given by the formula

$$\frac{n-1}{n} zp(1-p).$$

We are not going to show how this formula is derived; it is a purely formal mathematical problem, which has nothing to do with the principles of the theory. By inserting into this formula the values $n = 20$, $z = 100$, $p = 0.087$, we obtain the value

$$\frac{19}{20} \times 100 \times 0.087(1 - 0.087) = 7.55$$

for the expected dispersion. This is the average dispersion which we may expect to find by repeated drawings of 20 groups, each of 100 single tickets. Lexis compares this theoretical value with the empirical dispersion (in our case, 7.01) and sees in the ratio of these two values an appropriate measure of the degree of agreement of theory and observation. In the case under consideration, the ratio is $7.01/7.55 = 0.928$, and this is not very different from 1. According to Lexis, this result affords a confirmation of the hypothesis that the occurrence of the letter 'a' in a particular Latin text is approximately random, in the sense in which this word was used in our definition of a collective.

LEXIS'S THEORY AND THE LAWS OF LARGE NUMBERS

Lexis's Theory can be founded on the general Laws of Large Numbers for statistical functions which we discussed at the end of the last lecture. I can only give a rough idea of this derivation, because it is mainly of a mathematical character.

We start with two statistical functions of n observations, the mean D and the dispersion S^2. The first of these is found by adding up the

149

results of the n experiments and dividing the sum by n, the number of experiments. In our case $D = 174/20 = 8.7$. The dispersion S^2 is calculated by subtracting D from each of the n single empirical values and finding the arithmetic mean of the squares of these differences. In our example, S^2 was equal to $140.2/20 = 7.01$. We denoted the number of single observations in each of the n experiments by z. (In our example $n = 20$, $z = 100$.) We can now find the ratio $D/z = d$, and in our case $d = 8.7/100 = 0.087$. This can be called the average of the n relative frequencies, whereas D was the average of the n actual results. Previously, we have considered this number d as being practically equal to the unknown probability p of the letter 'a' and we used it above in this way in approximating the expected variance $\dfrac{n-1}{n} zp(1-p)$ by $\dfrac{n-1}{n} zd(1-d)$.

The more basic approach to Lexis's Theory lies in considering the ratio of the dispersion S^2 to this product $\dfrac{n-1}{n} zd(1-d)$ as a statistical function. This ratio, which except for the factor $(n-1)/n$, is known as *Lexis's Ratio*,

$$L = \frac{S^2}{zd(1-d)}$$

is, indeed, a statistical function, since it is derived from the two statistical functions S^2 and d, while z is a fixed number independent of the results of the experiment. Assuming that n is a sufficiently large number, two conclusions can be drawn from the Laws of Large Numbers applied to L.

First, the value of L observed in a sequence of n experiments on a collective whose distribution is known can be expected to differ only very slightly from its expected value calculated from this distribution. If the element in the collective under consideration is the observation of z identical and independent simple alternatives, it is found that the expected value of L is equal to 1. Thus, if n is a large number, we expect the value of L derived from n actual observations of this kind to be nearly 1.

Second, if the collective is not completely known, we are entitled to assume that the (unknown) theoretical value of L which corresponds to this collective is close to the observed value L. If this observed value L is not very different from 1, while the number n of observations is sufficiently large, we may presume that the collective under consideration is at least approximately one which leads to a theoretical value of L equal to 1. This consequence of the Second

Law of Large Numbers is of the greatest practical importance in the application of Lexis's Theory.

The reader may have noticed that in the first discussion of Lexis's ideas we have compared S^2 with

$$\frac{n-1}{n} zp(1-p),$$

whereas in this section the value of S^2 was divided by

$$zd(1-d).$$

The difference is explained by the fact that now we are considering only very large (strictly speaking, infinitely large) values of n, and with an indefinitely large value of n the ratio $(n-1)/n$ becomes practically equal to 1.

NORMAL AND NON-NORMAL DISPERSION

In the statistics of the occurrence of the letter 'a' in Cæsar's *De Bello Gallico*, we found a satisfactory agreement of the observed variance (7.01) with the theoretically expected one (7.55). In other words, Lexis's ratio was not very different from 1. The assumption that each experiment is a combination of z independent simple alternatives appears to be justified in this case.

Lexis's method is always applicable when each of the n numbers under consideration is the outcome of the observation of z simple alternatives. Examples of this kind are the statistics of deaths in a community of z individuals or the statistics of male births in different groups of population, e.g., in different towns of a country, or in a single town in different years. In general, the number of observations z is in those cases a more or less variable one. For simplicity we prefer, however, to ignore this complication and to assume that we compare sequences of observations of practically equal size.

It would be quite wrong to expect that in all such cases the results will be similar to those obtained in the example dealing with the occurrence of the letter 'a'. This and similar cases, in which a value of nearly 1 is found for Lexis's ratio, belong to the class of phenomena described as having 'a normal dispersion'. As we have seen, the occurrence of a normal dispersion permits definite conclusions with respect to the nature of the underlying collective.

We are now going to consider cases in which Lexis's ratio is considerably larger or smaller than 1. In these cases we speak of supernormal or subnormal dispersion. Our first problem is to find an explanation for the deviation of L from unity, i.e., to look for collectives with L-values other than 1.

SEX DISTRIBUTION OF INFANTS

The relative number of male and female births has been a favourite subject of statistical investigation. In the course of the 24 months of the years 1908 and 1909, 93,661 infants were born in Vienna;[8] the average was therefore 3903 births per month. Of these infants, 48,172 were boys, corresponding to an average of 2007 male births per month. The proportion of male births for the whole period is

$$48,172/93,661 = 0.51432.$$

For the 24 single months, this proportion varied from 0.4990 in March 1909, to 0.5275 in August of the same year. The following table contains the 24 monthly values:

$$0.5223, \ 0.5125, \ 0.5141, \ 0.5246, \ 0.5126, \ 0.5136,$$
$$0.5187, \ 0.5213, \ 0.5105, \ 0.5203, \ 0.5124, \ 0.5141,$$
$$0.5143, \ 0.5093, \ 0.4990, \ 0.5097, \ 0.5140, \ 0.5089,$$
$$0.5129, \ 0.5275, \ 0.5178, \ 0.5130, \ 0.5177, \ 0.5027.$$

The average of these 24 values is 0.51433; the dispersion, calculated according to the rules given above, is 0.000 0533.

Now, according to Lexis's theory, we ask for the value of the expected variance; this value is calculated from the distribution in the collective, the single elements of which are as follows: 24 repetitions of a group 3903 draws from an urn containing tickets, the proportion of those marked M (Male) being about 514 to 1000. The relevant formula is slightly different from the one used above because we are now considering, not the number of the events themselves (i.e., the number of male births), but their proportion. The result is that the factor z appears now in the denominator instead of the numerator of the fraction, and that the expected value of the variance is

$$\frac{n-1}{n} \times \frac{1}{z} p(1-p).$$

With $n = 24$, $z = 3903$, and $p = 0.514$, the numerical calculation leads to the value

$$\frac{23}{24} \cdot \frac{0.514 \times 0.486}{3903} = 0.000 \ 0613.$$

The ratio of the empirical dispersion to this last value (which we may think of as a theoretical or expected value of this dispersion) is

$$\frac{0.000\ 0533}{0.000\ 0613} = 0.869.$$

The actual dispersion is smaller than the theoretical one. In other investigations of the proportion of male births, a value of Lexis's ratio closer to 1 is obtained. We must therefore look for an explanation of the slightly subnormal dispersion found in this special case. It is, in fact, possible to give such an explanation by means of a more elaborate analysis based on the theory of probability. So far, we assumed a collective in which each element was represented by 24 groups of 3903 draws each from one and the same urn. These draws correspond to as many births occurring under identical conditions. In the new collective we still assume that we have 24 identical groups of 3903 draws each; but for each of these 3903 draws different urns are now used, i.e., urns with varying proportions of M's (= population groups with varying sex ratio). It follows from a known algebraical theorem,[9] that in this kind of collective the expected value of the dispersion is smaller than in that considered before. As it is very probable that the sex ratio of live births depends on race or social conditions or both, it must be expected that observations within a more or less mixed population will show a smaller dispersion than those within a perfectly homogeneous one. This hypothesis seems appropriate to explain the subnormal dispersion of the sex ratio in Vienna, at this time.

STATISTICS OF DEATHS WITH SUPERNORMAL DISPERSION

Examples showing a supernormal dispersion are much more frequent; here the dispersion found from the experimental data is much larger than the expected one. We may consider, for instance, the death statistics in Germany in the ten years 1877 to 1886,[10] during which period nothing out of the ordinary intervened to upset the general trend of life. The table on the next page shows the absolute number of yearly deaths together with the proportion of deaths per 1000 inhabitants:

A reader without knowledge of mathematical statistics may find the numbers in the last column amazingly uniform. Authors of an earlier period were full of astonishment at the apparently exceptional 'stability' of the conditions of human life. If, however, the dispersion of the values in the above table is calculated and compared with the expectation according to Lexis's Theory, the result is a very different one. The arithmetical mean of the ten numbers in the last column is 27.41 per 1000, or 0.02741. The ten deviations from the mean, squared (e.g., for the first year, $0.028 - 0.02741 = 0.00059$; $0.00059^2 = 0.000\,000\,3481$), summed, and divided by 10, give the value $0.000\,000\,0949$ for the dispersion. This number appears at first sight to be very small indeed.

If we now compute the corresponding expected value of the dispersion, we find that the number obtained is much smaller still. This is due of course to the very large value of z involved. We have to put $n = 10$, $p = d = 0.02741$, and z equal to the number of experiments in each of the ten groups of observations, i.e., the number of inhabitants of Germany, which was at that time about 45 millions. The calculation gives the result

$$\frac{9}{10} \times \frac{0.02741 \times 0.97259}{45,000,000} = 0.000\,000\,000\,533.$$

The ratio of the observed dispersion to this expected value is $949/5.33 = 177$; in other words, the actual dispersion of the yearly death frequencies is nearly 200 times the expected one. How is this to be explained?

Year	Total number of deaths	Number of deaths per 1000 inhabitants
1877	1,223,156	28.0
1878	1,228,607	27.8
1879	1,214,643	27.2
1880	1,241,126	27.5
1881	1,222,928	26.9
1882	1,244,006	27.2
1883	1,256,177	27.3
1884	1,271,859	27.4
1885	1,268,452	27.2
1886	1,302,103	27.6

SOLIDARITY OF CASES

Let us consider more closely the collective with which the yearly death roll is compared in Lexis's Theory. Its element is a tenfold repetition of a group of 45 million draws from an urn containing 274 black balls, and 9726 white balls, per every 10,000. If the correct analogy were to let every German draw once a year a ball

APPLICATIONS IN STATISTICS

deciding his life or death during the following year, the ball being replaced before the next trial so as to keep the proportions of the balls constant, then the dispersion of the death numbers would be 177 times smaller than it actually was. It is, however, obvious that this description of the game of life and death is very inadequate.

We know from everyday experience that there are phenomena that cause the death of a multitude of people, such as bad weather during a particular winter or summer month, an unfavourable economic condition within a district, or an epidemic. Therefore, we shall approach the real situation much more closely by assuming that instead of every single one of the 45 million inhabitants drawing his own lot from the urn independently of all the others, a smaller number of 'representatives' do so, each for a group of individuals. According to the formula of p. 149, the expected value of the variance increases in the same proportion as the number z of independent cases within a group decreases. Therefore, if we assume that a common lot is drawn for every group of 177 inhabitants of Germany, deciding on the life or death jointly for all of them, then there is complete agreement between theory and observation.

TESTING HYPOTHESES

Modern statistical methods, which play an important role today especially in England and America, group together a number of problems under the heading of 'Testing Hypotheses'. In this sense, Lexis's Theory tests the hypothesis that a certain sequence of observations can be considered as the outcome of a group of alternatives, i.e., experiments with constant probability p of success. The term refers, however, primarily, to 'hypotheses' which can be expressed numerically. The simplest instance of testing such an hypothesis is the problem which we discussed earlier and denoted as Bayes's problem.

Suppose we draw ten times from an urn which contains black and white balls. If we obtain seven black balls, we will readily adopt the hypothesis that this urn contains more black than white balls, i.e., that the probability of drawing black is greater than that of drawing white; in other words, that the probability p of drawing a black ball lies in the interval 0.5 to 1. We might even venture to restrict our hypothesis further, saying that p lies between 0.6 and 0.8, since the observed frequency is 0.7. The main question is: How is such an hypothesis to be justified, and are we able to indicate exact criteria for its adequacy?

155

Prior to making any calculations, we wish to mention two undeniable facts. First, under no circumstances can we draw from the given observations a *certain* conclusion concerning the unknown probability p. It could be that even with a very small value of p the first ten drawings might give seven black balls out of ten, even though such an event is quite 'improbable'. Second, it is clear that the numerical ratio 7/10 as such cannot be the only decisive instance for justifying the validity of our hypothetical assumption. If in 1000 experiments we obtain 700 black balls, the ratio is still 0.7, yet the hypothesis that the original probability p lies between 0.6 and 0.8 now has much better 'backing'.

According to our conception, the problem posed can be solved unequivocally and satisfactorily by means of Bayes's theory. Let us briefly review the solution of this problem. Given the numbers n and n_1 (in our present example 10 and 7), namely, the total number of all observations and the number of 'successful events', we want to know how great the probability is that the original probability p will lie between certain limits p_1 and p_2. It has been shown that the answer depends on the product of two factors. One factor is the familiar probability that, with a given p, n_1 black balls will appear in n experiments. This is given by the formula $w(n_1; p)$. The second factor is the generally unknown, initial or a priori probability of the particular value of p under consideration. Next, we have to form the product of these two factors for all values of p lying between the limits p_1 and p_2, and to sum up these products. Finally, according to the rule of partition, this sum is to be divided by the sum of such products for *all* possible p-values. This quotient gives us the desired probability of inference. We found that in this problem the influence of the factor representing the a priori probability tends to decrease more and more as the number of observations n increases. It therefore follows that if the initial or a priori probabilities of the possible values of p are really unknown, we *cannot draw any conclusions* so long as n is small. If, however, we perform a large number n of experiments, we can obtain meaningful results by making an arbitrary assumption about the a priori probability, e.g., that it is constant. This finding is in complete accord with common sense: if an event has occurred twice in three trials, we cannot conclude anything from this fact; if it occurs 2000 times in 3000 observations, we can draw fairly precise conclusions concerning the probability of the underlying value of p.

This same line of thought finds far-reaching applications in questions of statistics whose scope goes well beyond the original Bayes's problem. Consider n observations and a number n_1 derived from them in some way; then the probability w of the occurrence of n_1 is a function of n and n_1 and may further depend on some unknown quantity p. Our problem is to draw an inference regarding this unknown value of p from the known values of n and n_1. This problem can be solved only by means of Bayes's theory. For every value of p, in the interval in which we presume p lies, we form the product of w and the generally unknown, a priori probability of p. The sum of these products (divided by a certain constant) provides the probability of inference that p lies in the presumed interval. Again, we find, in most actually occurring applications, that as n gets larger the influence of the a priori probabilities becomes progressively smaller. It follows that if we have no information concerning the object of our observations, and the number of experiments n is not large, we cannot draw any conclusions; however, if n is sufficiently large, we shall obtain a good approximation with computations based on the assumption of a priori probabilities evenly distributed over all the possible values of the variable p.

R. A. FISHER'S 'LIKELIHOOD'

In 1921, the British statistician R. A. Fisher[11] attempted to found a new theory of inference; in this connexion, he introduced the term 'likelihood'. In common usage, 'likelihood' and 'probability' have the same meaning; Fisher, however, introduces the term in order to denote something different from probability. As he fails to give a definition for either word, i.e., he does not indicate how the value of either is to be determined in a given case, we can only try to derive the intended meaning by considering the context in which he uses these words.

Fisher starts with the premise that the initial probability $v(p)$ is unknown, at least in all practical instances, and that it may not even exist. In any case, he argues strongly that it should in no way enter into our considerations and our calculations. Given this point of view, we are left with only one factor—instead of two—for determining the probability of inference. If we denote again by w the probability of observing the number n_1 in n experiments, then, according to Fisher, this quantity w, which of course depends on the unknown p, is called the likelihood of p. Let us consider an example. It is easily seen (and can be deduced from the formula given on p. 149) that the probability of obtaining two successes in three experiments with the probability p for the single experiment is equal to $3p^2(1 - p)$. It

157

follows that, if $n = 3$, $n_1 = 2$, the 'likelihood' for $p = 1/2$ is equal to $3 \times 1/4 \times 1/2 = 3/8$ and the 'likelihood' for the value $p = 2/3$ is equal to $3 \times 4/9 \times 1/3 = 4/9$, i.e., somewhat larger.

As soon as he has introduced his new term, Fisher starts using it as completely synonymous with 'probability' in the everyday sense of the word. That is, he considers a value of p as more reliable if it has a greater likelihood, and he recommends in particular, as the 'best estimate' of p, the value which has the greatest likelihood. In our example of two successes in three trials, we can easily see that the greatest value of $3p^2(1 - p)$ is 4/9 and that therefore the corresponding value of $p = 2/3$ would represent the best estimate of p. (It is generally true that in the cases where the likelihood is expressed by the formula on p. 149 the value $p = n_1/n$ has the greatest likelihood.)

From our point of view, there is no doubt that likelihood is a correct measure of the probability of inference in two cases. First, where we have reason to assume that the unknown initial probabilities are uniformly distributed over all the possible values of p, or, in other words, that the initial probability is a constant. In that case, the quantity w is multiplied by the constant and the inferred probability of p is proportional to its likelihood. Second, if we know nothing about the a priori probabilities but, if the number of experiments n is very large, then we know that the influence of the initial probabilities is not very considerable and we may therefore consider them as uniform, in the sense of an approximation. Here again the probability of inference is, at least approximately, proportional to the likelihood. What meaning the likelihood could have in any case other than the two just described is inconceivable to me. I do not understand the many beautiful words used by Fisher and his followers in support of the likelihood theory. The main argument, namely, that p is not a variable but an 'unknown constant', does not mean anything to me. It is interesting to note that some philosophers have already begun to expound 'likelihood' as a new kind of probability which would not depend on relative frequencies.[12]

SMALL-SAMPLE THEORY

One might think that there is no great harm in introducing the word 'likelihood' for one of the factors of the product on which the probability of inference depends. In fact, unless one of the two cases just described should prevail (large number of experiments or approximately uniform initial distribution), we would expect that no one would even think of applying the theory. Unfortunately, this is not the case. Fisher emphatically avoids all reference to Bayes's solution of the problem of inference; this is for him a matter of principle. Therefore he cannot admit that, unless we have some information concerning the initial distribution, his notion of likelihood is applicable only in cases of a large number of experiments. Certain statisticians have found in this a welcome occasion to reach at last what seemed to them one of the goals of statistical theory, namely, to draw meaningful conclusions from a *small* number of observations. If in three experiments we have had two positive results, it is wonderfully satisfying to be able to make the statement that the likelihood of the unknown p having a value of 2/3 is 0.444. (Everyone secretly thinks that now he has a chance of 44.4% of being right in selecting the assumption $p = 2/3$.) This very idea, applied to more complicated problems, is at the basis of a system which is appropriately called Small-Sample Theory, namely, the theory of small groups of experiments. Let us illustrate it by a characteristic example.[13]

It is desired to test a new drug to see whether it induces longer sleep. The drug is given to ten patients and the result is that in the most favourable case sleep is prolonged by 3.7 hours; in the least favourable case, it is shortened by 1.6 hours. The average extension of sleep in all ten observations amounts to 0.75 hour. Now, without reference to any a priori knowledge, such as we might derive from the composition of the drug, the small-sample theory draws the conclusion that in the above case there is a likelihood of 0.888, i.e., almost 90%, that, on the average, sleep will be prolonged by the use of the drug under consideration. I do not think that any sensible doctor will have much confidence in this figure of 90%.

According to our way of thinking, if we have ten observations whose results oscillate between plus 3.7 and minus 1.6, we cannot draw any conclusions unless we include some a priori information, namely, some knowledge concerning the drug, gained independently of and in addition to our ten experiments. If it is impossible or too difficult to find a numerical expression for such a priori knowledge, we have no other recourse but to extend our sequence of observations to many hundreds or thousands of cases.

As a matter of fact, it almost seems that the heyday of the small-sample theory, after a rather short duration, is already past. There is little reference to it nowadays. We can only hope that statisticians will return to the use of the simple, lucid reasoning of Bayes's conceptions, and accord to the likelihood theory its proper role.

SOCIAL AND BIOLOGICAL STATISTICS

The field of application of statistical methods becomes wider as time goes on. The first field in which statistics found extensive use was that of social phenomena. Some people still believe economics and the social sciences to be the only legitimate domain for the application of statistical methods. The original meaning of the word 'statistics' is indeed probably 'the science of the state'. This part of statistics has long since found its undisputed place among the social sciences. If some exponents of social statistics maintain that it has nothing to do with the theory of probability, this is not to be taken too seriously. It is the same as if somebody were to say that it is possible to be a statesman without knowing the history of one's country, or to build bridges with no knowledge of statics. The classical achievements of social statistics were certainly due to a different outlook.

One of the founders of social statistics, Adolphe Quetelet,[14] has prefaced his fundamental work, *Social Physics*, by a detailed introduction dealing with the theory of probability, contributed by the astronomer Herschel. This book, which contains the famous concept of the 'average man', has influenced two generations of scientists. The subjects treated by Quetelet are often of a biological nature, and even include problems of medicine. The statistical treatment of biological phenomena, which is sometimes called biometry, nowadays plays a steadily growing part in the general science of life. English scientists have been especially active in its development. Far-reaching hopes of a possible improvement of the human race are often attached to the study of this subject; the term 'eugenics' has been specially designed to characterize this purpose of vital statistics. These high aims have not always been favourable to the soundness of the scientific basis of these investigations. A far sounder way has been followed in the development of a closely related subject, that of the science of heredity. The first statistical treatment of this subject was due to the Augustine monk Gregor Mendel (about 1870).[15]

MENDEL'S THEORY OF HEREDITY

Mendel recognized that the distribution of certain hereditary attributes in a number of organisms belonging to the same generation is similar to the distribution of attributes in a collective. Similarly, the inheritance of properties from generation to generation can be considered as a collective. With respect to a so-called 'Mendelian character', each individual possesses two 'genes', determining the possible alternatives. The colour of the flowers of peas is such a Mendelian character, the alternatives being red or white. The colour of the seeds of peas is also a Mendelian character, with the alternatives green or yellow. From the point of view of hereditary properties with respect to the colour of flowers, each pea plant belongs to one of the three types: white-white, red-white, or red-red, as determined by the corresponding genes. In the process of the formation of a new individual, a parent plant gives to each plant of the next generation one of its own genes, the other one coming from the second parent plant. The probability of the transfer is the same for both genes and is thus equal to 1/2.

Assuming that both parent plants belong to the type red-white, the probabilities of a plant in the next generation belonging to the groups red-red or white-white are 1/4 each. The probability of the type red-white is 1/2, since no distinction is made between the two combinations red-white and white-red. In this particular case the 'mixed' type is externally indistinguishable from one of the 'pure' types, which is called 'dominant'. Consequently the distribution of apparent attributes in the second generation is $(1/4)/(3/4) = 1/3$.

The actual observations of Mendel applied to the colour of seeds gave 2001 green peas and 6022 yellow peas in a heap of 8023 peas. Bateson found among 15,806 peas 3903 green and 11,903 yellow ones. The ratios are 1/3.01 in the first case and 1/3.05 in the second. Both these findings are in excellent agreement with the theory.

Far-reaching conclusions can be drawn from the simple assumptions of Mendel's theory. Although the underlying hypotheses cannot in general be tested directly, all important conclusions derived by the application of the theory of probability to the data of heredity have been confirmed most brilliantly by the numerous experiments of plant growers and animal breeders. This practical success is one of the most impressive illustrations of the usefulness of the calculus of probabilities founded on the concept of the collective.

The statistics of hereditary phenomena is nowadays a science widely developed and most successful in its application. It offers a number of interesting problems to the theory of probability and some of these are rather difficult. None of them is, however, of primary importance for the foundations of the theory.

INDUSTRIAL AND TECHNOLOGICAL STATISTICS

Industry offers some possibilities of application of the theory of probability which do not involve any fundamental difficulties either.[16]

161

Some of them belong to the field of the theory of errors, which will be discussed in some detail below. An example of this kind is provided in testing the uniformity of steel balls manufactured for ball bearings, where an accuracy in size to 0.01 mm is required. Also certain problems of a new character have arisen which are of great theoretical and practical interest. We may call them problems of Traffic Density.

The machinery of an electric power station must be chosen in accordance with statistical data concerning the frequency and intensity of the current consumption by its customers. The population of persons using the current can be considered as a collective. Attributes of a single element in this collective might be the period of the day when current is required and the duration and intensity of the consumption.

The problems arising from the construction of automatic telephone exchanges, and even switchboard problems belonging to the same group, are more interesting from a mathematical point of view. It would be impossible, or at least very wasteful, to build a telephone exchange so that all possible communications between any two subscribers could be established at the same time. The rational way is, starting from statistical observations and using certain probability assumptions verified by observation, to calculate the probabilities of different numbers of calls originating simultaneously, and to plan for combinations whose probability is not below a certain level. It is of course possible for highly improbable accumulations of calls to happen on some day. The engineers prefer the risk of a failure of the exchange in the case of such a highly improbable rush to an uneconomic increase in its size. This is an especially instructive example of the close relation between probability and relative frequency.[17]

AN EXAMPLE OF FAULTY STATISTICS

The picture which I' have drawn of the possibilities and results of statistical methods would be incomplete were I not also to say a few words on the erroneous and sometimes senseless theories which have been propounded in the name of statistics. Unfortunately, the number of such mistakes has been quite large, especially in medical literature. They are valuable as examples of the danger involved in each deviation from the firm principles of the theory of probability. In nearly all cases of this kind it is possible to recognize the source of error simply by inquiring a little more closely into the nature of the collectives to which the calculated distributions are supposed to correspond. I will discuss one example in some detail, because it has been published with the authority not only of a celebrated psychiatrist, but also of a well-known mathematician.[18]

In reciting certain Latin verses, a patient could not remember the word 'aliquis'. He was asked to give his nearest mental associations to this word. He first mentioned the word-division 'a-liquis'; then 'reliquiae', a 'liquid', and seven other expressions belonging, more or less clearly, to the complex 'blood-liquid'; finally a tenth association of a neutral character. These are the statistical data. They show that 9 out of 10 ideas associated by the patient with the word which he could not remember belonged to a certain complex of ideas. A calculation was now carried out, from which it was concluded that the probability of the lapse of memory being due to the displacement of the word 'aliquis' by the critical complex of ideas is a number differing from 1 by a fraction with 25 zeros after the decimal point! In other words, the authors conclude that the well-known explanation of this lapse of memory, which according to the psychoanalytical theory of Freud is due to a displacement, has this overwhelming probability of being true. More than that: it is actually hinted that the same immense probability, amounting to certainty in practice, can be ascribed to Freud's theory as a whole. We will overlook the obvious error of judgment in this last statement, since it is clear that even the most exact and certain verification of the predictions of a theory in *one single* case cannot be considered as a proof of its truth in general. We are interested here in the gross mistakes involved in the calculation of the probability relating to this one particular case.

The above-mentioned number with 25 zeros is obtained by the author in the following way. He estimates that about 1 in 1000 of all the 'ideas of an educated man' bears some relation to the 'critical' complex. He then proceeds to the solution of a problem of the Bernoulli type: the determination of the probability of drawing nine black balls in ten draws from an urn containing only one black ball per thousand balls. The answer is in fact a number of the order of 0.1^{26}, i.e., a decimal fraction with 25 zeros behind the point. The mathematical solution is correct, but what has this mathematical problem in common with the actual case under consideration?

CORRECTION

According to this solution, the outcome of the statistical experiment shows that the patient is forming a much larger number of associations belonging to the critical complex than an average educated man can be expected to make. However, in that case, the author's first task should obviously have been to determine experimentally how often ideas belonging to the critical complex occur among the associations normally provoked by the word 'liquis'. It is very improbable that an educated man, starting his associations with the syllables 'liquis', will not soon arrive at the words 'reliquiae' and 'liquid'. Once these words have been formed, it is also natural for him to remain for some time in this realm of ideas. The psychiatrist himself has obviously the same turn of mind, since he says that the separation of the word 'aliquis' into 'a' and 'liquis' already points in the direction of the critical complex. In any case, the initial collective has obviously been wrongly defined. The correct basis would be the number of ideas belonging to the critical complex associated by a 'normal' subject with the word 'liquis' and not the frequency of these 'critical' ideas among all the possible ideas of a normal individual.

The deduction of the final collective from the initial one is also wrong. In the language of our scheme of urns, we observed the relative frequency $a = n_1/n = 0.9$ of black balls in 10 draws. Let us assume the probability of drawing a black ball to be 1/2. The problem should then be formulated as follows: Under given conditions, what is the probability that the urn from which the draw '9 black balls out of 10' has been obtained contains a proportion of black balls greater than 50%? Or, in the language of psychology, what is the probability that the patient who has produced nine out of ten 'critical' associations has more than a normal inclination to produce associations of this kind? This problem is of the Bayes type and not of the Bernoulli kind, i.e., it belongs to problems leading to the Second rather than the First Law of Large Numbers. The probability asked for in this problem is about 0.95.

The derivation of this result involves the assumption of equal a priori probabilities for all the urns with different proportions of black balls. According to a proposition mentioned in the discussion of the Second Law of Large Numbers, the result would be independent of these a priori probabilities only if the number of experiments n was sufficiently large, whereas here $n = 10$ only.

Still a further uncertainty is introduced by our assumption of the value 1/2 for the probability of a 'critical' association with 'liquis' being formed by a 'normal' individual. All that we can say is that the result we obtained, i.e., the estimate 0.95 for the probability of Freud's explanation of this particular case, is a reasonable one, as compared with the abstruse assertion of a probability of 10^{-26} for explanations different from that suggested by Freud's theory.

We have thus learned that in order to treat this case correctly by means of the theory of probability it is necessary: (1) to take a different initial collective; (2) to make a plausible estimate of the probability for which we took the value 0.5, (3) to use Bayes's formula instead of Bernoulli's formula, and (4) to make a much larger number of tests so as to become independent of the unknown a priori distribution.

Actually, the whole method of investigation is probably not the most suitable one. It would be more realistic to assume that the word 'aliquis' provokes associations belonging to the critical complex in *all* individuals, and to investigate the relative frequency of the cases in which this word is forgotten by individuals under different psychological conditions. If such a statistical investigation is impossible, it would be better to abandon the statistical approach altogether rather than to force a conclusion for which there exists no statistical basis.

SOME RESULTS SUMMARIZED

I do not intend to add to the number of examples showing the application of the theory of probability to statistical problems. The fundamental problems have been made sufficiently clear by the examples we have discussed so far, and a systematic investigation of the various problems does not belong to the scope of these lectures. Some of the problems indicated above will be considered further in the next lecture, especially the phenomena of 'probability after-effects'. Before passing to a new subject, I will give the usual summary of the most important concepts which we have so far discussed in this lecture:

(1) Sequences of numbers obtained by statistical investigations can sometimes be immediately considered as collectives; in some other cases they can be reduced to a combination of a number of collectives.

(2) Arguments against the theory of probability based on the notion of the collective cannot be founded on existing statistical data. Neither Marbe's theory of 'statistical stabilization', nor the (opposite) theory of 'accumulation', suggested by Sterzinger, nor the supposed Law of Series provides valid arguments of this kind.

(3) The reduction of a series of statistical data to a corresponding collective can often be carried out by means of Lexis's 'Theory of Dispersion' and by more recent developments of this theory. This

is possible for the rather common type of problem where the statistical data are obtained from a series of simple alternatives.

(4) The Anglo-American statistical theory based on Fisher's notion of 'likelihood', which rejects Bayes's solution of the problem of inference, provides meaningful results in two instances: (a) if the assumption that the initial distribution is approximately uniform seems acceptable; (b) if the sequence of observations is long enough so that the influence of the initial probabilities has decreased. The so-called small-sample theory is to be completely rejected.

(5) If the concept of probability and the formulæ of the theory of probability are used without a clear understanding of the collectives involved, one may arrive at entirely misleading results.

DESCRIPTIVE STATISTICS

I have mentioned at the beginning of this lecture that a number of different definitions of the scope of statistics have been suggested. Many will call the subject discussed in this lecture 'Mathematical Statistics'. I would like to mention, however, that mathematical methods are also used in statistics in a way which shows no relation to the problems which we have discussed. Mathematical treatment is often applied to statistical data, without any probability considerations.

This kind of mathematical statistics can be conveniently called 'descriptive'. The word 'descriptive' is used here in its narrow sense, implying nothing but the formulation of results in mathematical terms, without attempting to include them in a more general logical system, based on the concept of probability.

The purpose of descriptive statistics is thus to characterize statistical results as exactly as possible, by calculating the values of certain characteristic functions. The simplest of these functions are the mean or average and the dispersion, which have already been discussed in the preceding sections.

These two quantities, average and dispersion, are in general not sufficient for a description of a series of statistical data and particularly for a comparison of various such series. Descriptive statistics has invented other methods of comparison, which usually consist in the calculation of certain additional statistical functions. A great number of statistical measures[19] of this kind have been suggested in the course of time; examples are the median or central value, various mean deviations, quartiles, deciles, etc. All these calculations together are not essentially more effective than the simple determination of the average and of the dispersion. A statistical function very useful in certain special cases is Gini's 'measure of disparity'.[20] Karl Pearson,[21] the founder of a great school of statistics in England, has tried another way of increasing the capacity of descriptive statistics, by defining certain typical distributions, in the hope of being able to reduce most of the distributions occurring in practice to one or another of these types. Mathematically the most perfect and most general of the descriptive methods suggested so far is that introduced by H. Bruns[22] in the development of his so-called 'Theory of Finite Populations'.

Starting from an idea suggested by the astronomer Bessel, Bruns showed how all statistical sequences can be described, with any desired degree of accuracy, by an infinite sequence of 'measuring numbers'. The first number in this systematically developed sequence is the mean; the second is the dispersion; the third measures the 'skewness', the fourth the 'excess', and so on. The Swedish astronomer Charlier has made valuable contributions to this method and shown how it can be adapted to other cases.

We have no intention of entering into a more detailed discussion of this subject, since the branch of mathematical statistics to which it belongs is the one least related to the theory of probability. All that is useful from our point of view is to know, in the interest of a general orientation on the subject, that all the different methods of descriptive statistics, including the Theory of Measuring Numbers, Pearson curves, and the expansion of Bruns and Charlier, are only methods for a preliminary treatment of the experimental data, by which we prepare them for the theoretical investigation.

FOUNDATIONS OF THE THEORY OF ERRORS

I want to add a few words on that branch of statistics based on the theory of probability which has found the most general acceptance and is the least controversial. It is equally important in the fields of statistics and of physics and therefore forms an appropriate link between this lecture and the one which follows. I speak of the so-called 'Theory of Errors'.

Most observations which depend on measurements, or, more generally, on any determination made on concrete objects, are liable to so-called 'errors of observation', namely, variations in the results obtained by the repeated measurement of the same quantity. Using the most exact methods available, with all possible precautions, we still obtain variations, for example, in repeated measurements of the distance between two fixed points on the surface of the earth. We

167

assume that this distance has a definite true, unchanging value, or at least a value which does not change over the period required for carrying out the measurements, and it follows that the different results obtained, with the possible exception of a single one, are all incorrect, that they all include errors of varying magnitude. There is no lack of possible sources of errors. When using the measuring rod we may even have made an allowance for the change in its length due to temperature variation, as we should. But still there is always possible an inaccuracy due to, say, direct sun rays, some not very well-studied change in the material of the rod, a slight bending, etc. Another source of error is the inexactness of the readings under the microscope, for two observers practically never make identical readings; there may be distortion through air currents, vibrations, elastic after-effects, imperfections of the optical apparatus, etc.

The problem of the existence or nonexistence of a 'true' value of every quantity belongs to the realm of epistomology and we need not be concerned with it here. The essential fact for us is that the results obtained in a sufficiently long series of measurements have all the properties of a collective. It is in general not possible to prove this directly by investigating the effect on the relative frequencies of different results if the number of observations is increased, or if place selections of different kinds are applied. In most of the applications of mathematical theories, the elements susceptible of an experimental test are consequences of the theory rather than of its premises. The same is true for the theory of errors.

The great mathematician, Karl Friedrich Gauss,[23] was one of the first to recognize the possibility of applying the theory of probability to the investigation of the errors of observation. He developed a method which is known under the name of the 'Method of Least Squares' and is widely used, particularly in geodesy and astronomy, but also in all other sciences having an observational basis. The method of least squares is based on a mathematical theorem which was first derived by Laplace;[24] its importance was not fully recognized until much later. This theorem is essentially a mathematical one, and I cannot explain it here in all its details, which are of interest only to the specialist. Laplace's proposition, however, is of a much too fundamental character to be completely passed over in these lectures. I shall attempt to explain its general meaning without entering into mathematical details.

We have already mentioned that errors of observation, or, more exactly, the mutual deviations of observations, are due to a great number of causes. The reader will find it plausible if I state that this very multiplicity of causes is the real basis for the existence of a general theory of errors. The individual sources of error may have differing regularities; all together, they produce a more or less uniform result. The general theorem which we are going to discuss is in fact founded on the idea that the resulting observed error is the sum of many small elementary errors.

GALTON'S BOARD

A very instructive model demonstrating the cumulative effect of a great number of independent causes of error is Galton's Board, which I am going to explain to you. It is a plain, slightly slanting board studded with 40 horizontal rows of pins. The distances between the pins in each row are all equal to 8 mm, and the distances between the rows are the same. An important feature is that each row is displaced horizontally with respect to the two neighbouring ones by 4 mm. The pins in the tenth row are in this way placed exactly underneath the middles of the spaces between the pins in the ninth row, and above the middles of the spaces in the eleventh row. The arrangement that the first few rows are shorter than the others is a practical detail and of no essential importance.

I now release a steel ball of 8 mm diameter, just above the middle pin of the top row. It can be deflected to the right or to the left. It then impinges upon one of the two middle pins of the second row, and again has to choose between the two possibilities of turning to the right or to the left. These and the following 'decisions' are made more quickly than I can describe them. When the ball has traversed all 40 rows, it has been deflected 40 times to the right or to the left out of its straight course.

We see that, in the present case, the ball has come to rest 4 places to the right of the middle of the board. This means that the number of deflections to the right has been greater by 8 than the number of deflections to the left, for a single deflection corresponds to 1/2 of the distance between two pins in a row, 40 deflections occurred altogether in the path of the ball, so 24 must have been to the right and 16 to the left.

The model represents 40 repetitions of an alternative with the attribute 'right' or 'left', or, in a more mathematical form, $+ 1/2$ or $- 1/2$. The whole process can be considered to be an element in a collective, its attribute being the resulting final deviation from the vertical course, that is, the sum of all the single deviations. In our case, this sum is $24 \times (+ 1/2) + 16 \times (- 1/2) = 12 - 8 = 4$.

169

NORMAL CURVE

Let us now assume that we have carried out a physical observation or measurement in which 40 independent causes of error were involved, each causing a positive or negative error which by an appropriate choice of unit can be made equal to $\pm 1/2$. The resulting error of the measurement as a whole is an exact equivalent of the resulting deviation of the ball from the straight course on Galton's Board. The distance by which the ball has been deflected from the middle corresponds to the total error of the measurement in the units chosen. Deviations to the right correspond to the resulting value being too large, those to the left correspond to this value being too small. What happens now if I make a large number of measurements instead of a single one? To see this, I have only to repeat the experiment on Galton's Board with a correspondingly large number of balls. Here are 400 balls, exactly equal in size and mass. One after another they find their way through the rows of pins. In a few minutes they have all arrived at the bottom of the board, and form vertical columns of different height in the different compartments provided for them. Each ball rests in the compartment whose distance from the middle corresponds to the sum of all the deviations made by it in the course of its passage. We see at once the whole result of the 400-fold repetition of the experiment, each consisting of 40 processes. The greatest number of balls lies in the compartments nearest the middle, corresponding to a deviation of not more than one or two units. The compartments situated symmetrically on both sides of the middle contain about the same number of balls. The whole distribution has a characteristic bell-shaped form. It illustrates immediately the distribution in the collective whose element is a 40-fold repetition of the original experiment.

Laplace was the first to calculate correctly the distribution of results obtained by the repetition of a great number of identical alternatives. This led him to the mathematical expression of the bell-shaped curve, which is nowadays usually known as the Normal or Gaussian curve. Fundamentally, the problem is identical with that solved by Bernoulli, which we have discussed in the special form in which the attribute was the number of 'heads' in the repeated tossing of a coin. This problem, however, and its answer apply only to any number n of repetitions, whereas Bernoulli's Law of Large Numbers as well as Laplace's result apply correctly only to infinitely large n. Bernoulli's Law of Large Numbers, in the form which Poisson has given to it, amounts to the statement of one significant property of the resulting distribution; Laplace has succeeded in finding a complete mathematical formula for this distribution. It is represented by the function e^{-x^2}. In other words, the ordinates of the normal curve decrease on both sides of the maximum in such a way that their (negative) logarithms are proportional to the squares of the distances from the middle.

Actually, there exist infinitely many normal curves characterized by different degrees of breadth, corresponding to different ratios of height and width. In the case of Galton's Board the degree of slenderness, or, as we usually say, the degree of precision, depends first of all on the number of rows of pins. It can be calculated by the application of the rules of addition and multiplication of probabilities, that is, by the fundamental operations of mixing and combination of collectives.

The general proposition underlying Gauss's theory of errors may now be formulated as follows: The normal curve, which is so clearly illustrated by the distribution of balls on Galton's Board, represents the distribution in all cases where a final collective is formed by combination of a very large number of initial collectives, the attribute in the final collective being the sum of the results in the initial collectives. The original collectives are not necessarily simple alternatives as they were in Bernoulli's problem. It is not even necessary for them to have the same attributes or the same distributions. The only conditions are that a very great number of collectives are combined and that the attributes are mixed in such a way that the final attribute is the sum of all the original ones. Under these conditions the final distribution is always represented by a normal curve.

We can easily understand how this proposition applies to the theory of errors. We assume only that in each observation a very large number of sources of error are concerned; it follows from this assumption, according to Laplace's Law, that the probability of a certain value x of the resulting error will be represented by the ordinate of an appropriate normal curve. This is the meaning of the basic concept of a 'Law of Errors', as represented by a normal curve, valid for all accidental errors of observation. This provides the basis for Gauss's theory of errors and his method of least squares. All formulas and methods of calculation which go under the name of the 'method of least squares' are consequences of this special form of the law of errors. In other words, they are all based on the assumption

171

that the errors of observation are caused by a great number of different causes. There are, of course, many other questions which I cannot discuss here. Some of them are even of basic interest, such as how to estimate the above-mentioned degree of precision of a given series of observations.

THE APPLICATION OF THE THEORY OF ERRORS

Gauss applied his theory mainly to geodetic and astronomical measurements. In these fields, the methods of calculation based on Gauss's theory, most of which were already known to Legendre, are an indispensable tool for all practical work. The theory is useful, however, in many other cases which have nothing to do with any 'errors' of observation, but rather with fluctuations of different kinds, such as variations among experimental results.

Take for example the case of measurements of the stature of a large number of individuals. If each result is expressed in units of 1/2 cm, practically no error of the individual observation will occur. It is, however, possible to consider the variations among the different results, i.e., in the heights of the persons measured as fluctuations caused by a number of different causes. The distribution of the different deviations from the mean value of all the measurements is found to correspond to a normal curve in this case as well. This is in accordance with our general postulate that a normal curve is always produced when the attribute of the final collective is the sum of a great number of independently variable 'accidental' attributes.

Actually, the results discussed at the end of the preceding lecture allow us to generalize the present proposition to include not only the sum or the average, but a great number of statistical functions. In this way, Gauss's Law applies to many biological and physical investigations where no errors of observation in the original sense of the expression are involved.

On the other hand, the field of application of the theory of errors should not be extended too far, as has sometimes been attempted. Not all the fluctuations occurring in the world follow the law of the normal curve. It would be quite wrong to assume that if this law does not hold the deviations from the mean value are not of an accidental character. Karl Pearson[25] and his school of biometricians were right in stressing that Gauss's Law is not 'the last word of wisdom' in statistics, and whilst in the work of the Pearsonian School we sometimes miss a somewhat deeper reasoning based on the theory of probability this should not blind us to the great progress in

172

biological statistics brought about by these investigations, with their freer conception of the nature of statistical distributions.

With these indications I should like to close my consideration of the application of the theory of probability to statistics. I shall return to certain questions related to the fundamental problems of the theory of errors in the last lecture when I deal with the interesting field of statistical physics and the problem of causality which is closely related to it.

Statistical Problems in Physics

ONE further aspect of the application of probability calculus remains to be discussed, that is the role played by the probability concept in the domain of theoretical physics. This development was initiated some fifty years ago when Boltzmann conceived the remarkable idea of giving a statistical interpretation to one of the most important propositions of theoretical physics, the Second Law of Thermodynamics. The significance of this idea, which bears on fundamental problems and concepts of modern science, is still increasing in our time. The basic points can be made clear even to an audience possessing no special knowledge of physics. For a first consideration of the problem it will in fact be better to forget about the actual content of the proposition and concentrate on its logical structure instead.

THE SECOND LAW OF THERMODYNAMICS

Classical thermodynamics, founded by Robert Meyer, J. Joule, and especially S. Carnot, has established that two functions play an important part in thermodynamic processes: energy and entropy. The Law of the Conservation of Energy, the so-called First Law of Thermodynamics, holds for the first of these functions; the so-called Second Law of Thermodynamics, which states that the entropy increases in every observable process, relates to the second. In consequence of Boltzmann's[1] famous investigations, started in 1866, this Second Law is now formulated as follows: It is extremely probable that the entropy increases in every observable process; a decrease in entropy is extremely improbable. Substituting in this proposition the definition of probability used throughout this book, we arrive at the following statement: If the 'same' physical process is repeated a very large number of times we may expect that, in the great majority of

174

cases, there will be an increase in entropy and only extremely rarely there will be a decrease in this function. This formulation is not complete; above all, we must specify what is meant by the repetition of the 'same' process; also, the relation between the value of the probability and the expected amount of increase or decrease in entropy must be explained, etc.

In any case, however, there can be no doubt that the Second Law of Thermodynamics deals with relative frequencies of observed results. We do not think that objections will be raised in this context to the interpretation of probability as relative frequency in an indefinitely long sequence of observations. The physicist Smoluchowski,[2] who made many valuable contributions to statistical physics, wrote in one of his papers: 'Mathematical probability is the relative frequency of the occurrence of certain important events'. This definition, although somewhat vague and containing an unnecessary restriction, is essentially our frequency definition of probability, as opposed to the classical definition based on equally likely cases. Physicists generally agree that probability should be interpreted as relative frequency, even though they do not usually enter into a discussion of this point.

DETERMINISM AND PROBABILITY

What was the new element in Boltzmann's statement that made it appear revolutionary? To answer this question let us consider what was for centuries our concept of the so-called 'laws of nature'. 'All bodies fall to earth with the same acceleration', said Galileo. 'Rays of light entering denser media are bent towards the normal to the surface' is part of Snell's law of refraction. 'Two small spheres carrying charges of the same sign repel one another' is a special formulation of Coulomb's law of electrostatic interaction. All these laws predict with absolute certainty the occurrence of a particular event given certain specified premises. These laws do not say that 'most bodies fall' or 'bodies fall almost always', the statements made are completely deterministic; there seems to be no possibility to relate them to the concept of probability. The conviction of scientists that all natural events have an unambiguously predetermined character is well expressed in Goethe's lines: 'Great, eternal, unchangeable laws prescribe the paths along which we all wander.'

The first instance of a deterministic explanation of a broad field of practical experience is found in Newton's Mechanics, as developed in his *Philosophiae naturalis principia mathematica* (1687). For nearly two hundred years thereafter all explanations of natural events followed the lines set down by Newton's principles; even Goethe, Newton's bitter adversary, was in agreement with him on this point. The most extreme formulation of Newton's determinism is to be found in Laplace's[3] idea of a 'mathematical demon', a spirit endowed with an unlimited ability for mathematical deduction who would be able to predict all future events in the world if at a certain moment he would know all the magnitudes characterizing its present state. Scientists have gradually tended to accept this point of view with respect to the inanimate world and possibly also with regard to plants and animals, though they may conceive that events of the past also exert an influence. In the opinion of the majority, and in accordance with the religious belief in free will, the human mind alone has freedom of action and of choice. Goethe's poem (Das Göttliche) quoted above continues:

'Man alone can achieve the impossible:
He distinguishes, chooses and judges.'

We are all convinced, if only instinctively, that man holds an exceptional position in the world and this belief has a decisive bearing on our point of view concerning the possible 'chance' occurrence of events.

We do not consider it strange that the number of deaths or of suicides in a certain country cannot be calculated in advance with the same exactness as, for instance, the predicted date of a lunar eclipse. The first two phenomena are directly or indirectly influenced by human free will as exercised in individual living habits, choice of residence, incurrence of risks, etc. We agree that all our decisions are influenced by external circumstances, but we do not consider this influence to be necessarily decisive. On the contrary, older authors, such as the English historian Buckle,[4] were astonished by the relative uniformity of statistical data which made it possible to predict future events of social life within reasonable limits.

Some interference on the part of free will can be seen even in games of chance, although these bear close resemblance to the processes of the inanimate world: We lift the dice box, shake it, etc., the winning lottery number is drawn by a child (or mayor), and to the last moment nothing is known as to which roll among thousands his hand will grasp. We have got used to the idea that processes of this kind are subject to the laws of chance, and that the theory of probability can be applied to them.

Should we then assume that purely mechanical or physical processes, the events of the inanimate world where no human hand interferes, are of a fundamentally different character? Surely, a mechanical loom carries out its complicated movements with unfailing regularity and yields products that are piece by piece exactly alike? If the kinetic theory of gases assumes that a gas consists of an immense number of invisibly small particles in a state of violent agitation, it would then seem to follow likewise (or at least so it appeared until recently) that the movement of each particle is completely and uniquely predetermined by the laws of mechanics. Our intellectual conscience seems to revolt against the idea that chance or 'laws of chance' could govern such processes.

CHANCE MECHANISMS

If, however, we submit this mental resistance to closer scrutiny, it proves to be nothing more than prejudice, as is so often the case. What is actually the difference between the 'purely mechanical' system of gas molecules in a closed vessel and the mechanism of a game of chance? Let us consider, for instance, a well known method used through the centuries to obtain 'pure' chance distributions—the lottery. The essential part of the game consists in the following steps: A sequence of numbers, say from 1 to 100,000 are printed on separate, equal slips of paper; these slips are rolled into small cylinders in as uniform a way as possible; all are placed into a large container which is kept in a circular or other motion by some appropriate mechanism; finally, a human hand arbitrarily seizes one of the cylinders. The most important element in this procedure is that the shaking of the container should mix the lots in a way which makes it impossible to follow the fate of any one of them. The method of the final drawing of a lot from the container is considered to be of no importance as long as the person who carries it out has no way of differentiating between the small cylinders. The best way to assure this last condition is to avoid the human element altogether and to provide some mechanism which would eject a lot through a funnel after a sufficient period of shaking the container. In fact, it has often been suggested that, in order to avoid all possibility of cheating, the whole process should be mechanized. There is no doubt that it is possible to construct a completely automatic 'lottery machine'. We imagine an electrically driven machine, a roll of paper is introduced, which in turn is imprinted, cut, rolled into small cylinders; the rolled lots are mixed and shaken in a container and

177

finally one of them, the winning one (or ones), is ejected through the exit funnel. This would be a fully mechanical device, complete in itself except for the motor transmission. And yet we are convinced that such a machine would not yield the same result each time but would, instead, follow the laws of chance, in the same way as a collective, so that the machine would eject, in a random manner, varying numbers for the winning lot.

Simpler chance mechanisms can also be devised. We mentioned in the previous chapter a game which consists in catching in a cup a steel ball rolling down an oblique board studded with pins, not unlike Galton's Board. A chance mechanism of this sort could be achieved, e.g., by fixing the cup in a certain position, to make it a game of pure chance rather than of skill, and adding a mechanical device for lifting the ball to the top of the board and releasing it there.

RANDOM FLUCTUATIONS

Once it has been accepted in principle that automatic mechanisms may produce results that vary according to the laws of chance, nothing prevents us from considering the behaviour of gas molecules in the same way. One can even go one step further in these considerations. We mentioned above the general opinion that a machine, such as the mechanical loom, will necessarily produce exactly identical pieces of fabric. Let us reconsider this. If two pieces of fabric produced by the same loom are compared by means of instruments more sensitive than the naked eye, small differences in the weave are sure to be discovered even if we leave aside faults due to irregularities in the thread. Larger or smaller deviations from the average width of the mesh, variations in the angles between the threads, these and similar irregularities occur even in the most accurately made fabrics. Actually a number of woven pieces will exhibit several different kinds of fluctuations; we can compare, for instance, the variations in the size of the mesh throughout one piece, or we can compare that variation for several pieces at a given place. Such fluctuations occur in all mechanical operations of even the most exact machines.

In speaking of the application of statistics to industry, I mentioned the fact that Gauss's theory of errors is of great help in the testing of steel balls for ball bearings. This example shows that the rational theory of collectives is applicable to and is actually applied in one of the most exact manufacturing processes of modern industry. According to the laws of mechanics, the steel balls produced by a completely automatic process should be completely determined. And yet this most exact mechanical process, which in the eyes of those who hold the deterministic point of view must be uniquely determined, is found, on closer inspection, to show chance variations. Is there actually any repetitive process which follows *exactly* the predictions of classical mechanics or of another similar physical theory? For instance, are the amplitudes of a simple pendulum, even under ideal conditions, all exactly equal? It is useless to argue about such questions. Even if we should find a mechanism which works without observable fluctuations it is only reasonable to expect that more sensitive testing instruments would reveal variations so far hidden. On the basis of our past experience we may moreover expect that these fluctuations, once revealed, will be found to have the properties of a collective and that, probably, the theory of errors will be applicable to them.

SMALL CAUSES AND LARGE EFFECTS

There is, however, an obvious difference between the above-described working of the 'lottery machine' and the fluctuations in the performance of a machine producing steel balls. In the latter case, it is possible to predict the *essential* properties of the product which the machine has been designed to manufacture. The laws of chance apply only to the small deviations from this main result. On the other hand, a 'chance mechanism' is built in such a way that *all that matters*, the essential result, is to be decided by chance. This formulation still contains something subjective in that we introduce the purpose of the mechanical process. If a machine designed to nail small boxes were geared to eject its nails by a mechanism similar to that used for ejecting the winning lot of the lottery machine, we would still not consider the former to be a 'chance mechanism' even though the mechanical setup would be the same in both instances.

From a purely mechanical point of view, regardless of the purpose of the machine, i.e., no matter whether we consider the meaningless ejection of nails for nailing small boxes, or the decisive selection of the winning lot in the lottery, the following property can be said to be an essential characteristic of a true 'chance mechanism': The results of repeated operations by the machine differ from each other by variations that are not of the order of small fluctuations or inaccuracies, but must be greater than a certain minimum deviation. In two lottery draws, the tickets winning the first prizes have numbers differing by not less than unity unless they are identical. On the contrary, in the case of mechanically produced ball bearings deviations may have any values however small in relation to the intended diameter while errors above a certain small amount almost never occur. We may bring out clearly the characteristic difference if we keep in mind that in all mechanical processes the initial stage is at least approximately predetermined. In the lottery machine this stage includes all operations through the rolling of the lots; during this stage only small variations can occur. We may therefore bring out the following essential attribute of a 'chance mechanism': Small, even infinitesimally small, initial deviations develop, in the course of the mechanical process, into large, finite differences.

Following M. v. Smoluchowski,[5] the physicist to whom we have already referred, we can characterize this essential property by the slogan 'small causes—large effects'. I will not go as far as Smoluchowski, and to a certain degree also Poincaré[6] before him, who considered this disparity between cause and effect to be the decisive property of *all* mass phenomena to which probability calculus might be applied. But as far as 'chance mechanisms' are concerned, namely automatic devices working in cycles and producing results that form a collective with a finite number of different attributes, Smoluchowski's characterization is surely appropriate. The lots fall into the container in the order in which they are printed and as long as the container is not agitated they remain in positions that would be *approximately* reproduced upon repetition of this operation. At the time the container begins to rotate, the existing small differences in the position of the lots produce initial conditions which vary slightly from one experiment to another. Under the influence of the rotation, these small initial variations develop into completely different final configurations, with the result that a different winning ticket is produced at the end of each process.

The case of Galton's Board is even clearer: a very slight difference in the point at which the steel ball touches the first pin determines the direction of the ball's first deviation. The same play repeats itself in the deviations occurring at the second and all subsequent rows of pins. It is reasonable to assume that the final result depends not only on the slight differences in the initial conditions, such as the velocity and direction with which the ball is released, but also on a number of outside influences coming into play during the process, e.g., air currents, vibrations of the board, etc. Each of these is in itself a very 'small cause', but together they decide the 'either-or' of the course taken by the ball, whether it will turn to the right or to the left of the next pin.

KINETIC THEORY OF GASES

The above considerations may readily be developed to permit us some insight into the phenomena which constitute the subject matter of statistical physics. The oldest example, which is in a certain sense representative of nearly all the later applications of statistics to physics, is given by the kinetic theory of the so-called perfect gas. The generally accepted conception is that space occupied by a gas is not uniformly filled with matter, as might appear at first sight, but that it contains tiny particles, atoms or molecules, moving to and fro in space with enormous velocities. The motion of a molecule is rectilinear, but its direction is changed again and again by collisions.

At a given moment, each molecule in the gas can be characterized by its velocity, or more exactly by the three components of its velocity with respect to some co-ordinate system. It appears to be appropriate, i.e., in accordance with experimental findings, to consider the molecules as elements of a collective, and to apply to this collective the rules of probability calculus. To arrive at an agreement with experience certain values must be assumed for the dimensions of the molecules, for their number in a given volume, and for their original distribution. Here are some figures which will be useful in our discussion:

We find good agreement with observation by assuming that under ordinary conditions a cubic millimetre of gas contains 30,000 billion (3×10^{16}) molecules.[7] The size of the molecules is such that three million of them lined up end to end occupy about 1 mm. The average velocity is of the order of several hundred metres per second. The average rectilinear free path of a molecule between two collisions is one ten-thousandth of a millimetre. Each molecule suffers about five thousand million (5×10^9) collisions per second. These figures show the appropriateness of such expressions as 'very small', and 'very many' when speaking of molecular phenomena. They also make clear why the rules of probability calculus, which actually presuppose an infinite number of events, apply so well to the practically infinite number of elements in the molecular collective.

As to the slogan, 'small causes, large effects', in what way does this essential characteristic of chance mechanisms apply to gaseous systems? To answer this question we must calculate an additional figure from the above data. The diameter of a molecule has been given as one three millionth of a millimetre, the mean free path as one ten-thousandth of a millimetre, i.e., 300 times as large. The geometrical conditions are thus the same as if balls 1 centimetre in diameter were colliding after a mean free path of 3 metres. It is clear that the very slightest deviation from the original direction of a ball will have a decisive influence on the result of the next collision. If, for instance, two balls of 1 centimetre diameter are placed a distance of 3 metres apart, and the first ball is set in motion along a line joining their centres, while the second is held at rest, then upon collision the first ball will rebound elastically and return to its original position. A deviation of about 9 minutes, i.e., 0.0004 of the circumference of the circle, from the original direction of the first ball is sufficient to make it rebound in a direction forming an angle of 45° with the original direction of its motion. Instead of returning to the point of departure, it will pass it at a distance of 3 metres. An initial change in direction of 16 minutes will produce a deflection of 90° after the collision. If the deviation is equal to 23 minutes, i.e., less than one-half of a degree, no collision will occur at all, the moving ball will pass the one at rest without touching it.

The conditions just described apply, strictly speaking, to Boltzmann's original model of a gas consisting of absolutely elastic spherical molecules. This hypothesis has been abandoned since and replaced by models of a more complicated character. The original theory is, however, very appropriate to demonstrate the large effect of small variations in the initial conditions on the further course of the motion of a molecule. The next stage of the molecular model, which today is likewise outdated in several respects, consists in assuming that each molecule consists of a larger or smaller number of electrons quickly revolving around one or several atomic nuclei. Such a model further enhances our impression that the result of a molecular collision is decisively affected by the slightest change in the original conditions of the system. Since it is assumed that the electrons revolve around the atomic nucleus with velocities much larger than that of the molecule as a whole, it is clear that the smallest change in the speed or direction of the molecular velocity will completely change the electronic configuration of the molecule at the moment of collision. Briefly speaking, small causes produce in this case larger consequences than in any game of chance.

A volume of gas containing a great number of molecules appears thus as a system not different in principle from the automatic lottery machine previously discussed, or from the Galton Board with an automatic device for handling the balls. The fact that we have a kinetic theory of gases which is built on the probability concept, and yields results in good agreement with experiment is no more remarkable than the applicability of probability calculus to games of chance.

We may even expect the application to gases to give especially good results because of the two conditions already mentioned, namely, the immensely large number of elements in the collective, and the extreme disparity between the small causes that influence the molecular collision and the large effects produced. At any rate, a statistical gas theory no more contradicts the causality principle than does any other statistical explanation of observed phenomena.

ORDER OF MAGNITUDE OF 'IMPROBABILITY'

The objection may be raised that a statistical formulation of a fundamental physical law involves practical difficulties apart from those of a logical nature. How is it possible to test the truth of a statistical proposition? Are we not compelled by experience to state that entropy increases in *all* real processes or, at least, that its increase can be expected with certainty if some specific conditions are satisfied? At first sight it would seem to be a question of actual fact whether this increase is a matter of probability of certainty. To throw light on this topic, I must discuss somewhat more closely the concept of entropy. The extremely large numbers of molecules in a gas volume and their extremely small size will play an important part in this discussion.

According to the kinetic theory,[8] a cubic centimetre of air contains 30 million billion (3×10^{19}) molecules, all in a state of rapid agitation. At any given moment the molecules are more or less uniformly distributed in this space. In the kinetic theory, entropy is considered to be essentially a measure of the *uniformity* of this distribution. Each of the 1000 tiny cubes, measuring one cubic millimetre, which compose the cubic centimetre contains, on the average, 30,000 billion molecules. According to Boltzmann, entropy has a larger value if the distribution is more nearly uniform, such that each of the 1000 compartments contains about the same number of molecules. Entropy has a smaller value if a marked unevenness exists, such that, for instance, one of the small compartments would contain only half the average number of molecules. Of course, a few more assumptions enter into this, e.g., it is assumed that for the whole volume under consideration, no marked differences exist in the physical state of the gas from one point to another, such as with respect to temperature or pressure, whereas such differences would be expected to exist in a larger volume. For an exact definition which would enable us to determine the value of entropy in a concrete instance, we would have to know other factors besides the distribution of the molecules in space, such as the distribution of the molecular velocities. All this, however, does not affect the basic logical structure of the problem.

Leaving the distribution of velocities aside, we can say that: according to all our experience, an abnormality in spatial distribution of the above-mentioned magnitude, i.e., a reduction to half of the number of molecules in one cubic millimetre, has *never* been observed. It is true that statistical theory does not completely exclude the possibility of such a distribution however abnormal it may be, it merely designates it as being most improbable. To get an idea of the degree of improbability attributed to such a distribution we may calculate its numerical value as given by probability calculus. The result of this calculation cannot be written down since it is a number with a hundred million zeros after the decimal point.

Such a small number is altogether unimaginable. Let us, therefore, take an example of a less extreme character. What is the probability of the number of molecules in one of the compartments exceeding the average number of 30,000 billion by a ten-thousandth of this value, i.e., by 3 billion? The result is still a number too long to be written down; it contains about 60 zeros after the decimal point. In this case, however, we can at least give a certain idea of the order of magnitude of this probability. It is roughly equal to the probability of winning ten times in succession in a lottery with one winning ticket in a million. Such a degree of improbability is almost indistinguishable from impossibility.

We thus recognize that experience cannot help us decide for or against the statistical conception in gas theory. From a practical standpoint, the consequences of the statistical theory do not differ from those of the deterministic theory.

CRITICISM OF THE GAS THEORY

I have considered the results of certain elementary calculations pertaining to the kinetic theory of gases in some detail because this theory is the oldest and most widely known application of the theory of probability to physics. The example is not quite satisfactory from several points of view. One of these has already been mentioned: it is the fact that the original simple spherical atoms and molecules of Clausius, Maxwell, and Boltzmann have since been replaced by other molecular models. First came, about twenty years ago, the concept of the atom as a kind of solar system, with 'planets' revolving around a central 'sun'. This model, due mainly to Rutherford and Niels Bohr, has in its turn suffered important modifications. The new and more general concept which had been anticipated by Ernst Mach, assumes that it is impossible to ascribe to the constituents of the atoms definite places at definite times, in the usual sense. Even more important from our point of view than the development of our conception of the atom is the fact that the logical foundations of the old gas theory have never been completely elucidated, especially in their relation to statistics. Boltzmann himself, in his gas theory, had not completely rejected the point of view of classical deterministic mechanics when he tried to reconcile his statistical explanation of the Second Law of Thermodynamics with a description of the mechanism of molecular collision along the orthodox lines of Newtonian mechanics. Boltzmann's point of view, which was shared by most of his contemporaries, was that statistical laws were valid for 'large scale' phenomena, i.e., observations made on a finite volume of gas, while the laws of classical mechanics governed 'small scale' processes, namely the motion of individual molecules. Ernst Mach rightly objected that the large scale behaviour required by the Second Law of Thermodynamics could never be deduced from mechanical laws valid for the small-scale phenomena.

Mach's criticism of Boltzmann's derivation of the Second Law of Thermodynamics is so important and unfortunately so often misrepresented that I feel obliged to say a few words about this problem. Mach[9] has shown that 'nothing analogous to an increase in entropy can be expected to occur in a completely elastic system of absolutely elastic molecules'; or in other words, that the Second Law of Thermodynamics cannot be derived from the mechanics of small elastic bodies. He says: 'How could an absolutely conservative system of elastic atoms possibly be made to behave as a system which tends towards a final state? No mathematical manipulations, however clever, can induce this.' Twenty or thirty years ago, when the kinetic theory of gases stood high in the esteem of physicists, these objections were disregarded. Actually, they aim at the very core of the matter. Indeed, by applying the laws of classical mechanics to the collision of small elastic bodies, nobody has ever succeeded—or could ever succeed—in deriving the thermodynamical properties of a gas as postulated by the Second Law; namely a preference for, and a tendency towards, 'less ordered' states, by that we mean states that exhibit a considerably uniform distribution of molecules, such as was exemplified in our previous mental experiment with the 1000 cubic millimetre compartments, each containing approximately the same number of molecules.

I may summarize the above arguments in the following way.

185

First, Boltzmann's idea of substituting statistical laws in certain cases for the usual deterministic laws of nature is in full agreement with our conception of games of chance and with the definition of probability based on this conception. Second, an empirical test to decide for or against the statistical interpretation of the entropy law is impossible because of the extremely small probability value ascribed by the statistical theory to the cases that the deterministic theory declares to be impossible. Third, a statistical conception applying to 'large scale' phenomena is irreconcilable with a deterministic conception on a 'small scale': statistical laws cannot be derived from the differential equations of classical mechanics. The new development of atom physics, which we shall briefly discuss at the end of this lecture has completely vindicated this point of view which I have held for many years.

We shall now discuss a few concrete problems of statistical physics before returning to general questions.

BROWNIAN MOTION

About a hundred years ago, the English botanist Brown observed under the microscope that certain organic liquids contain small particles moving to and fro in an incessantly agitated manner. It was discovered later that this so-called 'Brownian motion' is common to all sufficiently small particles suspended in a gas or in a liquid, and that it represents a mass phenomenon following the laws of probability calculus. Since we are only interested in the fundamental logical structure of this problem, we can simplify our conception by considering a two-dimensional scheme. We assume that the particles move in a zig-zag course in the horizontal plane, excluding any up or downward motion, or else, we may say that we consider only the projection of the true three-dimensional motion onto a horizontal plane.

Let us imagine that the bottom of a vessel is covered with square mesh graph paper that divides the plane into a great number, say N, small squares. We can now define the following collective: each Brownian particle is an element, the order number of the square in which it is found at a certain moment is its attribute. Experiments agree well with the assumption that the probability of finding a particle in a given square is the same for all squares, i.e., equal to $1/N$. In order to test this assumption, we can derive from the given initial collective a new one whose elements are observations of the momentary positions of a certain number n of particles; the attribute of each element is again the order number of the square in which it is found. By performing a 'mixing' operation on this collective we can obtain a collective whose elements are simple alternatives, e.g., square No. 25 contains 3 particles, or does not contain 3 particles. Next, we may take a large series of photomicrographs of the whole emulsion on a plate divided into N equal squares by a net of coordinates; we can now count with what relative frequency a given attribute appears, e.g., how often square No. 25 contains 3 particles. The agreement of statistical theory with observation is remarkably good in this case. In a series of experiments carried out by Svedberg[13] in which the mean concentration of particles was 1.54 per square the probability for finding exactly 3 particles in one designated square was calculated to be 0.130. The corresponding relative frequency in 518 observations was found to be 0.133 (namely 69 times in 518). Strictly speaking, in order to satisfy the requirement that the individual observations must be 'independent' from one another, it would be necessary to mix the emulsion thoroughly between counts, as is done in a lottery or in a game of cards. The results of experiments show, however, that an agreement of theory and observation is also obtained if the natural course of phenomena is not interrupted by mixing. This is a very remarkable result and we must discuss it somewhat more closely.

EVOLUTION OF PHENOMENA IN TIME

The main interest of physical statistics lies in fact not so much in the distribution of the phenomena in space, but rather in their succession in *time*. The same was true in the theory of gases, where we were mainly interested in the change of entropy in time. The usual way of observing the Brownian motion consists in fixing the attention on one square and counting the particles in this square, from second to second, or in other appropriate time intervals. How are the rules of the theory of probability to be applied to such a sequence of observations? Do the numbers of particles observed constitute the elements of a collective which can be derived from the previously described initial collective? Are we dealing here with a collective at all?

A simple consideration is sufficient to show that this is not the case. Let us observe the frequency with which a certain number of particles, say 7, appears in a given square. By repeating the observation on a very large number of time intervals, we can expect the relative frequency of the result 7 to approach a limiting value, thus satisfying the first condition of a collective. If, however, our observation covers not only the number of particles at a certain moment, but also the number in the same square in the immediately preceding moment, we will undoubtedly find that 7 follows more frequently after 6 or 8 than after, say, 1. The second property of a collective, that of randomness, is therefore absent in the sequence under consideration. The physical cause of this behaviour is obvious: During the small interval between two consecutive observations only a limited amount of motion occurs, so that small changes in the number of particles must be expected more often than large changes. The time sequence of observations of numbers of particles in a square is therefore not a collective in the sense of the theory of probability. We shall see, however, that it is possible to establish a relation between such a sequence and a collective. In the preceding lecture, we have already met sequences which, without forming collectives, could be reduced to one. It is comparatively simple to indicate a similar operation in the case of Brownian motion. To show this I will first use the example of a simple game of chance, the tossing of a coin.

PROBABILITY 'AFTER-EFFECTS'

Let us consider an alternative with the possible results 0 and 1 and assume that a sequence of observations thus represented by a row of zeros and ones possesses all the properties of a collective. Note, in particular, that in the long run, 1 will appear as often after 0 as after 1. Let us now calculate sums of pairs of adjacent results, adding the first figure to the second, the second to the third, the third to the fourth, etc. In this way the original sequence, say

$$1\ 0\ 1\ 1\ 1\ 0\ 1\ 0\ 0\ 1\ 1\ 0\ 0\ 0\ 1\ 0\ 1\ \ldots$$

is converted into the following:

$$1\ 1\ 2\ 2\ 1\ 1\ 1\ 0\ 1\ 1\ 2\ 1\ 0\ 0\ 1\ 1\ 1\ \ldots$$

which contains three different figures, 0, 1, and 2. It can be shown that for this sequence, just as for the original one, limiting values of relative frequencies exist for each of the three attributes 0, 1, 2. On the other hand, the randomness of the original sequence has disappeared: 2 can never follow 0, 0 can never follow 2. If, e.g., the fifth sum were 2, this would mean that the fifth and sixth figures in the original sequences are both 1's; whereas, if the sixth sum were 0, this would mean that both the sixth and the seventh figures in the original sequence are 0's. The two facts are inconsistent since the sixth figure in the original sequence cannot be both 1 and 0 at the same time.

The above is a typical example of a sequence of numbers derived from a true collective, which has itself been obtained from an observation of chance phenomena. The derived sequence still satisfies the first condition of a collective but has no longer the characteristic property of complete randomness. The mathematical analysis of the situation is as follows:

We consider a collective whose elements are groups of three consecutive elements of the original collective, the first, second, and third; the fourth, fifth, and sixth; and so on. Each triad can be characterized by two sums, the sum of its first two numbers and the sum of its second and third numbers. The usual rules of the theory of probability permit us to calculate the probabilities of the different combinations of these two sums, e.g., 0–1, 1–0, 2–1, 1–1, etc.; according to what was said before, the combinations 0–2 and 2–0 are impossible. Another collective of the same kind can be obtained by forming triads of the elements 2 to 4, 5 to 7, etc.; and a third by forming the triads of the elements 3 to 5, 6 to 8, etc. These three collectives have the same distributions, i.e., the probability of the simultaneous occurrence of two sums, such as 2–1, is the same in each of them. This probability is therefore also the limit of the relative frequency of the combination 2–1 within the sequence of sums of any two consecutive elements of the original collective. The limit of the frequency of the combination 2–1 so determined will generally differ from that of the combination 1–1 or 0–1 and this is why in the derived sequence we cannot speak simply of the 'probability of the attribute 1', in the sense of our usual definition.

M. v. Smoluchowski,[10] who was the first to study this phenomenon in connexion with Brownian motion, has given it the very suggestive name of probability 'after-effect'. This, like other suggestive names, has sometimes given rise to erroneous interpretations. The expression is meant to describe a sequence that results from a random process, but in which the attribute of an element is nevertheless affected by the attribute of the preceding element (or elements). The important thing for us is that the phenomenon of probability 'after-effect' of sequences which do not themselves form collectives can be explained on the basis of our theory of collectives which *do* satisfy the condition of randomness.

RESIDENCE TIME AND ITS PREDICTION

Let us return to the topic of the Brownian motion. The analogy between this problem and the one which we just considered is established in the following manner. Imagine that the time of observation is divided into a sequence of short elementary time intervals separated by 'instants of time' at which the observations are made. We assume that the changes of position of the particles occur by jumps during the elementary time intervals. Each co-ordinate of a given particle at a given time will be considered as the sum of the co-ordinate at the initial instant and of the subsequent co-ordinate increment. If we consider two successive instants of time, each co-ordinate of a given particle at the latter instant will be thought of as the sum of that at the former instant plus the change that took place in the intervening time interval. For the treatment of the time-development of Brownian motion we therefore require two initial collectives. The first gives the probabilities for the presence of a particle at a given spot at the initial instant. The element of the second is the change in the co-ordinates of a particle occurring during an elementary time interval. Thus, the attribute within this collective consists of positive or negative increments of the two co-ordinates of the particle. In other words, we assume as given the probabilities of any specified *change* in position of a particle during an elementary time interval. Probabilities of this kind, which may be called 'transition probabilities', are of decisive importance not only in the treatment of Brownian motion but likewise in many other statistical problems of physics.

We use the two collectives just introduced to construct a new one by observing the n particles through k instants of time. The attribute in this collective consists of the $2\ nk$ numbers representing the co-ordinates of the n particles at the k moments of observation. The next step consists in mixing the attributes of this last collective in order to obtain a collective whose attribute is the proportion x/k of time points at which the number of particles present within a given square (i.e., having some specified co-ordinates) equals, say, three. In other words, we can calculate the probability that a given space 'optically isolated' under a microscope will be found to contain exactly 3 particles at say, 1 % of all time points observed. The result of this computation is very characteristic and we proceed to describe it briefly.

The ratio x/k is called the relative residence time of a specified number of particles within a given square. In calculating the probabilities for the various possible values of this ratio (all of which lie between zero and one), we assume, in accordance with physical reality, that the number of particles n and the number of squares N are very large and that the period of observation is likewise considerable so that k is large too. Then we find that among all possible values for the relative residence time x/k there is *one* with a probability overwhelmingly larger than all others, in fact practically equal to unity. I have previously mentioned that from the first initial collective we can calculate the probability that a given square (we spoke of No. 25) contains at a given moment exactly three particles. Naturally, this probability is a proper fraction and its value in Svedberg's experiment was found to be $p = 0.130$. The result of further calculations, in which the second collective, i.e., the 'transition probabilities', are used shows that, after a sufficiently long period of observation, the probability of the ratio $x/k = 0.130$ is practically equal to unity. In order to be able to describe this result in a short and precise way, it will be convenient to introduce certain new terms.

The value of the probability $p = 0.130$ was derived from the simple assumption that the probability of finding a particle in one of the N squares is $1/N$. The arguments here used belong essentially to the field of combinatorial analysis. We may, therefore, call this probability briefly the 'combinatorial probability' of three particles in a certain square. The fraction x/k (where x is the number of instants of time at which three particles were observed at a given place, and k is the total number of instants within the period of observation) has been called by us the relative residence time. This residence time refers, of course, to the presence of any three particles, that is, to the 'number three', and not to three specified particles. We may now briefly formulate the result of our calculation as follows: It is practically certain that the relative residence time of the 'number three' within a given square will be approximately equal to the combinatorial probability of three particles in this square. The probability of a relative residence time which would deviate considerably from the combinatorial probability is negligible.

This proposition exhibits the form characteristic of all statements made in the statistics of physics. No deterministic prediction is made about the future course of a physical process, as is done in the usual 'causal' branches of physics, e.g., in classical mechanics; instead, a certain numerically characterized development is said to possess an overwhelming probability, so that the probability of any other development is exceedingly small. We have noted these same characteristics in reference to the kinetic theory of gases. The new element here is that we have now shown how definite, concrete predictions concerning the course of a physical process can be derived by means of the theory of probability without any use of mechanics or any other 'deterministic' science, and even though the corresponding

191

sequence of observations does not form a collective, since the property of complete randomness is lacking. This is the typical structure of the propositions of statistical physics.

Let us add two brief comments. First, to arrive at the above result, we had to know the distribution in the initial collective, which gave the probabilities of the larger or smaller 'jumps' of the particles. We found, however, that if the numbers n, N, k are sufficiently large, the final result was practically independent of the numerical values of these probabilities. Second, a similar remark applies to the probability after-effects: There is an overwhelmingly large probability for the relative frequency of a given *succession of particle numbers* in a square (e.g., first 2, then 3 particles) to be nearly equal to the combinatorial probability of this particular succession; however, the relative frequency, say for '3 after 2', is different from that for '3 after 1' or '3 after 5', etc.

We have thus indicated how the time sequence of numbers of particles in a square can be rationally treated by means of the theory of probability, although it does not form a collective.[11]

ENTROPY THEOREM AND MARKOFF CHAINS

Let us return to the fundamental idea of Boltzmann's gas theory which we shall now be able to understand better. In his conception, entropy is a characteristic of the gas, and we are interested in its variation in time. Even if we do not adopt a deterministic point of view, we have no right to assume that the successive values of a physical variable form a collective, i.e., that they are distributed at random like the results of a game of chance. In general, the application of the theory of probability to variations in time of physical properties is similar to that illustrated in the example of Brownian motion, which is why the above investigation is of such great importance for us.

To make the discussion sufficiently general, we have to replace the notion of the after-effect by the broader one of linkage of events, or, as it is often called, of 'Markoff chains'.[12] The essential point is that again the main role is played by a collective, whose elements are transitions, displacements, or jumps of the system from one state to another. These transition probabilities are basic in the analysis of the problem. The numerical values of these transition probabilities are usually unimportant, only their general type matters; this fact, however, is irrelevant for the logical understanding of the theory.

The physical characteristic that we considered in studying

Brownian motion was the number of particles in an optically isolated field. Correspondingly, for a volume of gas, we may consider the distribution of the molecules in the different 'compartments' of the 'velocity space' (or, more precisely, the so-called 'phase-space' which includes both space and velocity co-ordinates). Again, each distribution possesses a certain combinatorial probability, calculated according to the rules of addition and multiplication of probabilities. A detailed study, based on a consideration of the transition probabilities, leads to a result that parallels that in the case of Brownian motion: We may expect with very great probability that the different distributions occurring in the natural time succession will appear with relative frequencies which are approximately equal to the corresponding combinatorial probabilities. And this holds even though this succession does not exhibit complete randomness. Instead of the term combinatorial probability, the term 'thermodynamical probability' is often used here.

The combinatorial calculations show that certain distributions of molecules possess an overwhelming thermodynamical probability as compared with all others. This result, together with our above finding, leads to the conclusion that a volume of gas, left to itself, will almost always be found in the state of the most probable distribution. Furthermore, if at some moment the system is in a state differing considerably from the most probable one, then it will almost always at the next moment be found to have changed in the direction towards the most probable state. These distributions, which claim almost the whole probability and therefore also almost the whole time, are called Maxwell-Boltzmann distributions, because, prior to Boltzmann, the English physicist Maxwell had already begun to investigate them.

Boltzmann's decisive and new idea was to recognize that the thermodynamical probability, or rather its logarithm, measures the entropy. The new formulation of the entropy law is thus: A less probable state is almost always followed by a more probable one. This is equivalent to the proposition: The entropy of a system left to itself almost always increases. The word 'almost', a symbol of indeterminism, cannot be omitted from these formulations.

We cannot go beyond these brief remarks. In order to enter further into the problems of gas theory, it would be necessary to introduce various thermodynamical concepts and corresponding mathematical complications. Let us therefore return to the simpler example of Brownian motion and illustrate the results of the theory by means of some concrete observations.

The degree to which the theory of Brownian motion can be tested is limited by the very nature of the experiments. It is impossible to perform so many sequences of observation that one could expect to obtain a result of such enormous improbability as that which we indicated in the case of kinetic theory. Rather, we may expect that a single sufficiently prolonged sequence of experiments will yield those results to which the theory ascribes an overwhelming probability. Practically speaking, this means that the observed frequency of a certain number of particles is being compared with the combinatorial probability calculated for this number. We may mention in passing that this is also the explanation for the fact which we noted earlier, that, even without mixing the emulsion between measurements good agreement is obtained between the counted frequencies and the corresponding combinatorial probabilities. The extent of this agreement will be illustrated by the following example taken from investigations on Brownian motion made by the Swedish physicist Svedberg.

Svedberg counted the number of particles in an optically isolated volume of a colloidal solution of gold at intervals of 2 seconds. In 518 counts, he found the following relative frequencies:

No. of Particles	Occurrence	Rel. Frequency	Probability
0	112	0.216	0.212
1	168	0.325	0.328
2	130	0.251	0.253
3	69	0.133	0.130
4	32	0.062	0.050
5	5	0.010	0.016
6	1	0.002	0.004
7	1	0.002	0.001

The average number of particles was $a = 1.54$. The values of the probability p for 0, 1, 2, . . . 7 particles computed from this average value a by means of probability calculus are shown in the last column and are seen to be in good agreement with the observed relative frequencies. The theory predicts (see p. 191) with a probability of almost 1 that the relative frequency of the occurrence of 3 particles will be close to 0.130. The corresponding value of the observed frequency in the one and only experimental sequence under consideration was 0.133.

It is interesting to investigate also the probability after-effects in this same example. Our table shows that the concentration '1' was observed 168 times. In 4 of these 168 cases, no immediately following observation was recorded; among the remaining 164 cases, 40 were followed immediately by the value 0, 55 by the value 1, 40 by the value 2, 17 by the value 3, etc. The relative frequencies of the values 0, 1, 2, 3 immediately following the value 1 were thus $40/164 = 0.246$, and similarly 0.336, 0.246, and 0.104 respectively. If we now calculate in the same way the relative frequencies of the values 0, 1, 2, 3 immediately following the value 3, we find 0.087, 0.334, 0.319, and 0.189. Comparing the two sets of results, we see that in the latter the frequency of a 0 is much smaller, that of a 3 much bigger than in the previous sequence which referred to the value '1'. This illustrates again the lack of randomness by showing that this sequence (consisting of the observation of successive numbers of particles) is not insensitive to place selection and is therefore not a collective.

It is nevertheless possible, as I have previously explained, to apply to this situation the rules of the theory of probability founded on the notion of the collective. It is possible, for instance, to calculate for certain pairs of successive particle numbers those relative frequencies which are expected to occur with an overwhelming probability. Omitting details of the calculation, we find in this way, e.g., 0.246 for the pair 1–0 and 0.116 for the pair 3–0, compared to the observed values of 0.246 and 0.087 respectively, while the probability of the value 0 by itself was given in the above table as 0.212. Since the total number of observations in the above series was not very large, the agreement between theory and experiment must be considered a good one.

RADIOACTIVITY[14]

Another physical phenomenon which has been found to be well suited to the application of the laws of probability calculus is the emission of rays by radioactive substances. Like Brownian motion, this is a phenomenon which can actually be observed by means of a relatively simple experimental apparatus, where no hypothetical concepts are involved that are as violently contested by different schools of thought, as the atoms and molecules of the kinetic theory of gases. The radioactive decay of a substance, say radium, is accompanied by the emission of so-called α-particles, minute constituents of the atoms which carry a positive charge. When incident on a screen covered with an appropriate sensitive substance, each α-particle causes a short flash of light called a scintillation. The time intervals between successive scintillations have a random character.

195

A very simple representation of the whole process can be given by assuming that a very large number of constituent parts of the radioactive body—and we might as well call them atoms—are at every moment (more exactly, at very short time intervals) faced with the alternative: decay or conservation, with respective probabilities p and q.

The same problem in a game of chance would be this: A die is thrown at short regular intervals, and the interval between successive appearances of the result '6' is noted. Our initial collective has thus the simple attributes '6' or 'not 6'. By a k-fold combination we derive a new collective whose elements are combinations of k throws of the initial sequence, and whose attributes are the combinations of the k results '6' or 'not 6'. Next, by a mixing operation we obtain a collective with the same elements, but with a new two-dimensional attribute. This consists of the number z of throws which precede the first 6, and the number x of throws from the first to the second 6, so that $x - 1$ is the number of the result 'not 6' between the two 6's. By making k sufficiently large, we may disregard the case of less than two 6's in the whole experiment.

The probability of any given combination of z and x can be calculated by means of the usual rules. If we denote by p the probability of the result '6', and by $q = 1 - p$ that of the result 'not 6', the required probability is $q^2 p q^{x-1} p = p^2 q^{x-1} q^z$. Mixing again, by adding all cases irrespective of the z-value but with the same value of x, we obtain the probability of the interval between the first and the second '6' being equal to x. This probability is given by the series $p^2 q^{x-1}(1 + q + q^2 + q^3 + \ldots)$. This series is, strictly speaking, not an infinite one but ends with the $(k - x)$th term, $k - x$ being the largest possible z value. By choosing a sufficiently large k, we are, however, entitled to substitute for the sum in parentheses the corresponding sum with an infinite number of terms, i.e., the geometrical sequence $1 + q + q^2 + q^3 + \ldots$, and this sum is known to be $1/(1 - q) = 1/p$. The probability of an interval x between two 6's is thus $p q^{x-1}$; i.e., p for $x = 1$; pq for $x = 2$; pq^2 for $x = 3$; etc.

We are now able to calculate the expected value of the time interval in question by multiplying in the usual way each value with its probability, and adding the products. We obtain in this way the result

$$p + 2pq + 3pq^2 + \ldots = p(1 + 2q + 3q^2 + \ldots).$$

Algebra teaches that the sum in parentheses equals $1/(1 - q)^2 = 1/p^2$. Thus the expected value of the time interval between the first and the

196

second '6', using an unbiased die with $p = 1/6$, equals $1/p = 6$. This result is surely not surprising and those who have not had enough knowledge or patience to follow the above calculation will accept it without protest. I have brought in these computations merely to show once again how this problem can be answered unambiguously by our methods of operating with collectives.

The same value $1/p$ is obtained if we calculate the expected interval between the second and the third, or the third and the fourth '6', and so on. The mean time interval between any two consecutive results '6' is thus seen to be equal to $1/p$, the unit of time being the interval between two consecutive throws.

PREDICTION OF TIME INTERVALS

In trying to apply these considerations to the case of radioactive radiation, the first difficulty that arises is that we do not know either the values of p and q or the length of time between observations. Moreover, we must assume that this time interval is so extremely small that no apparatus will be sensitive enough to measure it accurately. Nevertheless, the following experiment may be used to test the theory:

We begin by calculating the probability that no more than three time units (i.e., 1, 2, or 3) elapse between two consecutive scintillations. According to the addition rule of probabilities, the result is:

$$p + pq + pq^2 = p(1 + q + q^2) = p\frac{1 - q^3}{1 - q} = 1 - q^3.$$

The probability of the opposite result, namely that there will be more than three time units between two successive scintillations is therefore equal to $q^3 = (1 - p)^3$. Taking instead of three an arbitrary number n, we obtain the probability $(1 - p)^n$ that the interval between two successive scintillations will be greater than n time units. It can be shown that if p is a very small and n a very large number, $(1 - p)^n$ is practically equal to $(1/e)^{np}$, where e is the so-called base of natural logarithms whose value is 2.718, so that $1/e = 0.3679$. This value taken to the (np)th power gives the probability of intervals greater than n time units occurring between two successive scintillations. Let us compare this result with observation.

We can observe successive scintillations for a certain length of time and note the time interval in seconds between the first and the last one. Dividing this time by the number of intervals counted, we obtain the mean length of the time interval between two scintillations.

197

Let it be, for example, 5 seconds. If the theory is correct and the period of observation sufficiently long, the observed interval of 5 seconds must be nearly equal to the corresponding expected value which we have calculated to be $1/p$. The time unit is thus $5p$ seconds, and the intervals 'longer than n time elements' are those that last more than $5np$ seconds. Let us assume $np = 2$; then the intervals exceeding $5np$ seconds are those longer than 10 seconds; intervals longer than 15 seconds correspond to $np = 3$, and so on. All we need do to test the theory is to count the number of intervals exceeding 5, 10, 15, . . . seconds, to divide these numbers by the total number of intervals observed, and compare these relative frequencies with the probabilities calculated above, i.e., with the successive powers of 0.3679.

In a more general way, we can say that if a is the observed mean length of time interval between two scintillations ($a = 5$ in the preceding example), the computed probability of intervals longer than na is given by the nth power of the number 0.3679.

MARSDEN'S AND BARRATT'S EXPERIMENTS[15]

A series of experiments reported by E. Marsden and T. Barratt included 7563 intervals between successive scintillations. The observation lasted altogether 14,595 seconds. The mean length of an interval was, therefore, $a = 1.930$ seconds. The following table contains the observed frequencies of the intervals exceeding a certain length (1 to 9 seconds). The next column contains the same values divided by 7563, i.e., the relative frequencies; the last one the corresponding theoretical probabilities. Since $1/a$ is 0.518, we have first to calculate $0.3679^{0.518}$. This gives 0.596. We have then to take the successive powers of 0.596, from 1 to 9, given in the last column.

Intervals larger than	Number	Relative Frequency	Calculated Probability
0 secs.	7563	1.000	1.000
1 ,,	4457	0.590	0.596
2 ,,	2694	0.356	0.355
3 ,,	1579	0.209	0.211
4 ,,	921	0.122	0.126
5 ,,	532	0.070	0.075
6 ,,	326	0.043	0.045
7 ,,	196	0.026	0.027
8 ,,	110	0.015	0.016
9 ,,	68	0.009	0.009

The agreement between the two last columns in the table is very good. Nevertheless, I do not deny that investigations must be continued and developed in many different directions before we shall understand the details of these phenomena; many questions in this field are still unanswered. Our main interest, however, lies in the general form of the statements which can be made concerning the evolution in time of physical phenomena by means of the theory of probability. In this regard, I should like to add the following remarks.

Our theory certainly does *not* imply that if in 14,595 seconds of observation a total of 7563 intervals between scintillations were recorded, it then follows that 21.1 % (1596 out of 7563) will have a length of more than three seconds. All we know is that 0.211 is the probability of this particular result. Combining this result with the First Law of Large Numbers, we can state that the probability that approximately 21.1 % of the 7563 intervals will have a duration of more than 3 seconds is almost equal to 1. Or, more precisely: if sequences of 7563 observations are repeated a great number of times, then the vast majority of them will contain approximately 1596 time intervals of more than 3 seconds. The form of this proposition corresponds exactly to that which we derived in the case of Brownian motion. In this latter case, we could say that if the (long) sequence of observations concerning the particles present in a specified square will be repeated a very large number of times, the vast majority of these sequences will give for the relative frequency of the occurrence of a certain number of particles in the square values nearly identical with the combinatorial probability for this number. It is essential to remember that the theory predicts not the exact result of a single sequence of observations but the outcome of the great majority of identical experiments (each experiment consisting of a large sequence of observations), repeated a very large number of times.

RECENT DEVELOPMENT IN THE THEORY OF GASES

Once more I return to questions of the kinetic theory of gases which formed the starting point of this lecture. I have said before that Boltzmann's first assumptions, which form the original framework of his statistical interpretation of the entropy theorem, have since been repeatedly altered. One of the primary aims in remodelling the theory was to provide a satisfactory explanation for the behaviour of gases at extremely low temperatures. In this region certain deviations from the normal gas laws occur which are described as the 'degeneration' of gases.

The various statistical theories which have been put forward in this connexion may be briefly described as follows: In different theories different initial distributions of probabilities are assumed. We know that such distributions have to be known in any probability problem. The object of all the theories is to assume the initial probabilities in such a way that for the medium and high temperatures the theory leads to agreement with the older theory of Clausius-Boltzmann, and, for the temperatures near the absolute zero, it reproduces a decrease in specific heat paralleling that which is actually observed.

The usual form of a hypothesis concerning the initial distributions consists in assuming that certain ranges of the attributes or certain regions in the space of attributes are equally probable. Variations in the initial distributions of probability are obtained by modifying the regions in the space of attributes which are assumed to be equally probable. Clearly, the objections against the use of the concept of 'uniform distributions' that we made when discussing the foundations of the theory do not apply in this case. Any distribution may be reduced to a uniform one by choosing appropriate co-ordinates and appropriate ranges of attributes. But it is important to realize that by so doing, one is making a special hypothesis in each case, and not using a principle given a priori. The following discussion is very instructive in this respect.

DEGENERATION OF GASES: ELECTRON THEORY OF METALS[16]

The classical theory of ideal gases was founded on the assumption of equal probabilities for all values of the velocity of a molecule, where the equal probabilities are defined as follows: Let us consider the three components of a velocity as the three co-ordinates of a point in a rectangular system. Each point in the space of these three co-ordinates is a point in the velocity space (which is the 'attribute space' for our initial collective) and corresponds to a certain possible velocity of the molecule. The assumption of the classical theory is that equal probabilities are assigned to equal volumes in this velocity space. We will call each element of volume in the velocity space a possible 'position' or 'place' of the molecule. If we now consider a collective whose elements are distributions of a certain number n of molecules over m positions in the velocity space, it follows that all possible n^m distributions have the same probability. For example, imagine two molecules A and B, and three different positions a, b, c.

The number of different distributions is 9, since each of the three positions of A, namely Aa, Ab, Ac can be combined with each of B. According to the classical theory, all these distributions have the same probability, $1/9$.

A new theory, first suggested by the Indian physicist Bose, and developed by Einstein, chooses another assumption regarding the equal probabilities. Instead of considering single molecules and as-suming that each molecule can occupy all positions in the velocity space with equal probability, the new theory starts with the concept of 'repartition'. This is given by the *number* of molecules at each place of the velocity space, without paying attention to the individual molecules. From this point of view, only six 'repartitions' are possible for two molecules on three places, namely, both molecules may be together at a, at b, or at c, or they may be separated, one at a and one at b, one at a and one at c, or one at b and one at c. According to the Bose-Einstein theory, each of these six cases has the same probability, $1/6$. In the classical theory, each of the first three possi-bilities would have the probability of $1/9$, each of the other three, however, $2/9$, because, in assuming individual molecules, each of the last three possibilities can be realized in two different ways: A can be in a, and B in b, or vice versa, B can be in a, and A in b.

The Italian physicist Fermi advanced still another hypothesis. He postulated that only such distributions are possible—and possess equal probabilities—in which all molecules occupy different places. In our example of two molecules and three positions, there would only be three possibilities, each having the probability $1/3$; i.e., one molecule in a and one in b; one in a and one in c; one in b and one in c.

In testing these and other hypotheses it is assumed, according to Boltzmann's entropy theorem, that the probability of the state of a gas is a measure of its entropy, and the object of the investigation is to find which theory best approximates the actually observed dependence of entropy on temperature and mass. Fermi's hypothesis was found especially useful in the study of free electrons in metals, i.e., in calculating the consequences of the assumption that the electric conductivity of metals is due to electrons which move to and fro between the atoms of the metal as in a strongly rarified gas.

It is not our purpose to discuss these problems here in more detail and still less to express any opinion about their relative merits. In as far as the foundations of the theory of probability are concerned, we may say that the gas theory with all its modern developments fits easily into the general scheme which we found typical for all statistical

201

theories: We have to calculate the distribution in derived collectives from given distributions in initial ones. If, as in the gas theory, the initial distribution cannot be derived directly from experiments, the assumptions must be tested by comparing the results of appropriate calculations with corresponding experimental data.

QUANTUM THEORY

Just as in the gas theory, the theory of probability plays an important role in the quantum theory. This far-reaching development started in 1899 with a fundamental discovery due to Max Planck.[17] The atomic concept of matter had been developed much earlier, mainly on the basis of chemical evidence. The assumption of an atomic structure of electricity was put forward towards the end of the last century as an unavoidable consequence of experimental results and the hypothesis was confirmed that all electrical charges consist of integral multiples of elementary charges, the elementary charge being that carried by a single electron. Despite these precedents, Planck's assumption of the atomic structure of thermal emission was a daring new idea; the whole modern atomic theory of light has developed from it.

Planck's original assumption was made in order to explain a certain observable dependence of thermal emissivity on temperature. Planck supposed that the emission of thermal energy by a body was the result of the action of a large number of ultramicroscopic 'oscillators'. The characteristic property of these oscillators is that they absorb and emit energy only in quantities that are multiples of a unit of energy called a 'quantum'. The quantum is directly proportional to the frequency of an oscillator. If, as is usually done, we denote by v the frequency, i.e., the number of oscillations per second of a given oscillator, then the unit of energy for this oscillator equals hv, where h denotes Planck's universal constant. In the metric system,

$$h = 6.55 \times 10^{-27} \text{ gcm}^2/\text{sec}^2.$$

Consequently, the energy of the oscillator emitting yellow light with a frequency of about $v = 5 \times 10^{14}$, is equal to $hv = 3.3 \times 10^{-12}$ erg. (One erg is the amount of work needed to lift one milligramme through one cm.)

According to this theory, the emission of monochromatic light, that is light with one single frequency, v, occurs in the following way: We have a large number n of oscillators at this frequency each containing 0, 1, 2, 3 or more quanta. *We consider these oscillators as*

202

the elements of a collective with the number of quanta as its attributes.
We must assume a certain initial distribution in this collective, i.e.,
probabilities for oscillators with 1, 2, 3, . . . quanta. The simplest
assumption is that for an oscillator all possible numbers of quanta
are equally probable and this leads to acceptable results. From the
above collective we can, e.g., derive a collective such that each of its
elements consists of n oscillators possessing together a certain total
energy E. The attribute within this collective consists of the distribution or 'repartition' of the possible numbers of quanta over the n
oscillators, i.e., how many oscillators possess one quantum, how
many have two quanta, etc. One of the various possible distributions is the most probable one. If n is sufficiently large, a Law of
Large Numbers permits us to expect with great probability that the
actual distribution at an arbitrary moment will almost equal
the most probable one. (We see that the case is similar to that of the
Maxwell-Boltzmann distribution in the theory of gases.) Using the
known thermodynamical relations between temperature and energy,
we can calculate the energy of radiation in relation to the temperature, a problem which, as we previously mentioned, formed the
starting point of Planck's investigations. The law thus obtained,
which is known as Planck's Law, is confirmed by very exact
observations.

From the point of view of the principles of the theory of probability, the same reasonings and deductions can be applied to the
above problems as to those in the kinetic theory of gases. I may
therefore refer the reader back to these discussions especially as regards the role of transition probabilities in the study of the evolution
of phenomena in time.

I should mention that Planck's original assumption was further
developed by Albert Einstein in 1905. Einstein introduced the concept of quanta of light with energy unit equal to $h\nu$ and conceive
light as being composed of atoms or 'photons'. A statistical theory
of light can be derived from this assumption. It contains, however,
certain elements which did not occur in the statistical theory of gases,
and in order to explain this new stage in the development of the statistical theory of quanta, I must first deal with a more general problem.

STATISTICS AND CAUSALITY

When I spoke of Brownian motion and of radioactivity, I indicated the essential difference between a physical theory based on
the usual notion of causality and a statistical theory. Deterministic

theories claim to predict the occurrence of events with certainty and precision from the initial data, while the statistical theory merely states what is expected to occur in the majority of cases in a long sequence of observations. At the beginning of this chapter, we mentioned that the transition from the deterministic to the statistical conception of certain physical phenomena was mainly due to Boltzmann's new interpretation of the classical law of entropy, first suggested about 1870. At that time, this meant breaking with the most cherished conceptions that had served physical science so well for over two centuries and had also penetrated into the minds of laymen. However, we are now living in a period of rapid development of physical theory; Boltzmann's gas theory and similar concepts, such as the theory of Brownian motion, the theory of radioactivity, and Planck's theory of oscillators, are now called 'classical statistics', and a new system, the so-called quantum statistics, has developed which goes much further in revolutionizing the conceptions of physics. Before beginning the discussion of this last point on my programme, I must add a few remarks on the relation between classical statistics and the so-called law of causality.

It is neither possible nor necessary to discuss here all the problems usually associated with the term 'law of causality'. One more reason to dispense with a full discussion is that Philipp Frank[18] published in 1932 an excellent detailed monograph on this subject. We shall briefly consider the only point of interest to us, namely, what is meant today by the causal explanation of natural phenomena. For this purpose, we need not trace this concept back to its roots in pre-scientific philosophy. Rather, we may consider it to have originated simultaneously with what we call today 'classical' science, namely with Newton's fundamental work, *Philosophiae naturalis principia mathematica*. In his mechanics of particles, Newton gave us the first example of a causal description of observable phenomena. For a long time thereafter the meaning of the word 'explanation' was to remain identical with that of 'causal description' in Newton's sense. A wide variety of phenomena were shown by Newton to appear as logical consequences of a small number of simple fundamental laws or axioms. What was the most important tool used in constructing this great edifice?

CAUSAL EXPLANATION IN NEWTON'S SENSE[19]

It was no accident that Newton, the founder of classical mechanics, was also one of the creators of differential and integral calculus. The use of calculus in deriving consequences from fundamental axioms is inseparable from Newtonian mechanics and from classical physics in general. The typical procedure is as follows: Axioms or fundamental laws are expressed in the form of so-called differential equations which state relations between certain physical variables in very small elements of space and time.

For example, in mechanics such a relation connects the velocity and position of a body at a given moment to its velocity and position at an immediately preceding moment. The mathematical process of integration then serves to derive from this differential law the changes in velocity and position of the body in a finite time interval.

Integration, however, can only derive from the position and the velocity of the body at a certain moment, its position and velocity at a later moment. This imposes a considerable limitation on our desire for a causal explanation. We are not told *why* the earth is revolving today, in a certain sense, with a velocity of 29.6 km/sec, nor why it is at a distance of about 23,400 earth radii from the sun, unless we accept the following as an answer, i.e., as a causal explanation: 'because one month ago it had a certain velocity and occupied a certain position'. Indeed, it is essentially a question of habit to see in such a consequence of Newton's mechanics a causal explanation of the corresponding phenomenon. The educated layman who hears for the first time that classical mechanics does nothing more than calculate the velocity at a given moment from that at a previous moment will surely call this a description rather than an explanation. Physicists were, however, distressed and annoyed when Kirchhoff, in 1874, advanced this very conception previously put forward by Ernst Mach, namely, that mechanics merely provides us with a systematic description of motions in the simplest possible terms.

We shall understand this point more clearly if we remember the historical development of the problem. In Kepler's work we can see the kind of questions originally raised in connexion with the study of the motion of celestial bodies. At school we learned of Kepler's three laws which were discovered as the result of painstaking observations of the planets; these laws, we were taught, were given a unified explanation through Newtonian mechanics. The textbooks, however, do not say that Kepler also stated many other laws which bear no relation to Newtonian mechanics. He postulated, for instance, that the axes of the orbits of the five major planets are proportional to the sides of the five Platonic bodies (tetrahedron, cube, octahedron, etc.) inscribed into one and the same sphere. It is not essential whether this assertion is correct or not. What is important for us is to note that Newtonian mechanics cannot be meaningfully related to this question. The initial position of a planet, that is, its distance from the sun at a given moment, is required by Newtonian mechanics as one of the input data needed in the calculations. Only by exercising a strong intellectual resignation with respect to a number of the questions which were raised in Kepler's time concerning the solar system are we able to say that these problems have been solved by Newton. All the developments that followed Newton's theory, including the celebrated theory of relativity, did not change this situation.

THE LIMITATIONS OF NEWTONIAN MECHANICS

Someone may protest against my last statements and argue that mechanics did not stop at the description of the current state of the motion of the stars but, beyond this, permits conclusions concerning the origins of the solar system and so approaches problems of the type discussed above. We may, for example, assume that the solar system was at first a single ball of liquid matter, as suggested by Laplace. We can further suppose that the formation of the celestial bodies from this mass occurred in accordance with the laws of Newtonian mechanics. The final state of the system must then follow from its original condition, and this final state would provide the initial conditions for the now stabilized motion of the celestial bodies. At this point, however, another intrinsic difficulty of Newtonian mechanics becomes apparent, and this is even more important than the first. I have not mentioned it before, in order to avoid undue complication in the first discussion.

In addition to the two classes of variables, position co-ordinates and velocities, the differential equations of mechanics, as well as the laws obtained by their integration contain certain magnitudes called 'forces'. In order to reach conclusions of the type described above by way of differential equations, we must know how these forces depend on the position and the velocity of the bodies in question. Thus, to describe the evolution of the solar system by means of Newtonian mechanics, we must know the laws governing the forces that operate in this process. If we say that Newtonian mechanics explains or describes in simple terms the motions of celestial bodies, it is because only one simple type of force, or rather one single law of force, occurs in all the corresponding equations. All astronomical motions (with a single exception) could be interpreted satisfactorily by means of the one assumption: that bodies attract each other with

206

a force inversely proportional to the square of their distance (Newton's universal gravitational force). One of the most momentous steps towards a more unified understanding of nature was Newton's proof that an apple falling from a tree and the revolution of the moon around the earth are described by equations containing one and the same assumption concerning the acting force. In such a case, we may indeed be inclined to speak of a causal explanation of the phenomena, i.e., their reduction to a common cause, 'gravitation'. Further development showed that even most of the details of the motions of the celestial bodies can be derived by the application of the law of gravitation with all its implications. This law, however, does not help us at all to understand the history of the solar system before it attained the comparatively stationary state in which we find it now.

In speaking of Newtonian mechanics, we do not only refer to astronomical theory. The motions of all material bodies on the earth are correctly described by the differential equations of Newtonian mechanics provided that appropriate expressions for the forces are introduced into the equations. For example, the theory works perfectly if we deal with complicated systems such as the motions of the different parts of a steam engine. All that is needed is the knowledge of the forces coming into play in the cylinder as a consequence of the thermodynamic process.

Let us, however, consider a mechanical process, such as the motion of a great number of uniform steel balls on Galton's Board. It appears at first sight self-evident that Newtonian mechanics must be valid in this case as well. What assumptions, however, are we to make about the forces? If we assume no force but that of gravitation, the essential result of the experiment will remain unaccounted for. In each collision of a ball with a pin, a very small force component decides about the direction of the ball to the right or to the left of that pin. This small component may, perhaps, be due to air currents that are unavoidable on account of the differences in pressure and temperature in different parts of the room. To describe adequately the movement of balls on Galton's Board by means of Newtonian mechanics, it might thus be necessary to know exactly the nature of the air currents in the room. Our desire for causality will, however, remain unsatisfied if we make an assumption concerning the air currents in order to account for the observed facts in a satisfactory way: we will wish to inquire further into the causes of these currents, i.e., we will wish to apply the equations of Newtonian mechanics to them as well, and therefore, we will want to know the laws of forces causing the air currents. We shall, however, not feel satisfied until we have found a *simple* assumption from which all the observed phenomena of this kind can be derived. Then at last can we feel that we have given a causal explanation of the phenomena under investigation.

SIMPLICITY AS A CRITERION OF CAUSALITY

Thus we find on closer consideration that a causal explanation of natural phenomena does not merely imply the existence of the typical pattern of classical mechanics (or classical physics in general), i.e., a system of differential equations which can be integrated given certain initial data, but that the additional criterion of a certain simplicity of all the assumptions is required. An instructive example is found in the very theory of planetary motions. It was discovered that a certain factor in the motion of Mercury could not be explained by means of the gravitational interaction between this planet and the other known members of the solar system. It would, of course, have been possible, without violating the mechanical equations, to account for this effect (a slow precession of Mercury's elliptical orbit) by assuming the action of other forces. These could even be considered as gravitational forces by postulating the existence of great masses of dust in the neighbourhood of the sun which would act as appropriate invisible sources of gravitation. However, the desire for causality was not satisfied by such hypotheses. Only when Einstein showed that by changing our methods of measuring times and velocities, the precession of the perihelion of Mercury could be quantitatively accounted for without assuming new forces, did most intelligent physicists accept this view, although it required intellectual sacrifices in other respects. From this example, we can see that the postulates of classical physics, which make all events appear as uniquely determined, satisfy our desire for causality and are accepted as a causal explanation of nature only if this is achieved by means of sufficiently *simple* assumptions concerning the forces introduced in the basic equations.

This condition of simplicity is not satisfied and cannot be satisfied in the case of those physical phenomena to which we apply the probability concept. It is a pure illusion to think that the motion of the balls on Galton's Board can be given a 'causal' explanation by means of the differential equations of classical mechanics, because, from a formal point of view, these equations can be said to determine unambiguously the course taken by a ball. And, if we try to arrive at a better understanding of the motions by applying more and more exact experimental methods, we shall soon find ourselves in new difficulties. Close microscopic examination of the air currents or of the surface of the solid bodies will merely reveal everywhere processes similar to that of Brownian motion. Instead of finding something which, from the point of view of classical physics, is more simple than the original phenomenon, we arrive at phenomena that are more and more complicated (from a deterministic point of view). Actually, it is well known today that the precision of all measurements, even of the simple measurement of length, is limited by the existence of Brownian motion. We shall discuss later further difficulties in exact measurements which are of an even more fundamental nature.

All this shows that the customary and suggestive concepts of classical physics are not applicable beyond certain limits.

GIVING UP THE CONCEPT OF CAUSALITY

At this point, there appears a new theory, the theory of probability, which shows a way by which definite conclusions concerning the course of certain natural phenomena—namely those which we have defined as mass phenomena—can be derived by means of logically unassailable arguments from simple assumptions about certain initial distributions. The statements of this statistical theory are not in disagreement with those of the deterministic theory, they do not even compete with them since they are of another form; they only state what will occur in the majority of cases, in a great number of identical experiments.

The empirical usefulness of the theory, i.e., the confirmation of its predictions by observations, has been established beyond doubt in many different fields. The intellectual resistance which stands, or stood until recently, in the way of a general acceptance of statistical theories is of a psychological rather than of a logical nature.

We certainly ought not to underestimate the influence exercised on our intellect by the century-old habit of a deterministic concept of nature. However, and I have discussed this in some detail elsewhere,[20] in the last few decades we have entered upon an epoch characterized by the development of speculative science on an unprecedented scale. As a result our whole scientific outlook has been broadened and fully transformed. The first great event of this epoch was Einstein's theory of relativity which stirred up interest far beyond the community of physicists. And yet, however great a task it was for

209

us to modify our deep-rooted habits of thinking, our old-established, near-sacred conceptions of time and space, all this appears almost insignificant beside the revolutionary upheaval created by some other ideas of modern physics. Using the tools of probability and statistics, these new concepts have led us from the kinetic theory of gases to quantum theory and wave mechanics. It now appears inevitable that we must abandon another cherished notion that has its origin in everyday life and pre-scientific thought and has been elevated to the rank of an eternal category of thought by overly zealous philosophers: the naïve concept of causality.

THE LAW OF CAUSALITY[21]

If we consider the vague and varied formulations which have been given by leading philosophers to the law of causality, we come to the conclusion that it is not at all easy, perhaps hardly possible, to contradict this 'law'. The first edition of Kant's *Critique of Pure Reason* says: 'All that happens (begins to be) involves the existence of something before it from which it follows according to a rule'. In the second edition, this formulation is replaced by: 'All changes occur in accordance with the law of cause and effect'. It would be perfectly possible to adapt any system of rules, including those of statistics, to these general conditions and this would amount essentially to deciding in each given case what is meant by 'change' and 'occurrence', by 'cause' and 'effect'.

When Galileo discovered the law of inertia through his observations, the notion that a continuous displacement can occur without a continuously acting cause was surely in contradiction with the concept of causality of his time. However, when generations of physicists found the law of inertia and its consequences to be a most convenient basis for the systematic description of the phenomena of motion, the philosophers yielded. In fact, Schopenhauer and many others went so far as to state that the law of inertia was an unavoidable logical consequence of the law of causality and held that only changes in velocity needed a cause but not changes in position. If the physicists had discovered that the causes influencing motion, the so-called 'forces', are proportional to the third derivative of the co-ordinate with respect to time instead of to the second, the philosophers would have declared the law that 'a body left to itself moves along a parabola' to be in agreement with, or even 'an unavoidable consequence of', the principle of causality.

Similarly, we can find formulations of statistical propositions that are in accord with the 'law of cause and effect'. If a 'double-six' appears on the average once in 36 casts of two dice, we can say that the 'cause' of this regularity is the fact that each die falls equally often on each of its six sides. All deviations from the frequency 1/36 can be said to be 'caused' by the 'bias' of the dice. When the balls rolling down on Galton's Board form the bell-shaped curve, we might say that the 'cause' of this result is that if a ball hits a pin repeatedly it will turn equally often to the right as to the left. Such statements may be countered by saying that no cause can be indicated for the single result in either the game of dice or Galton's Board, or that fluctuations in short sequences of observations are not traceable to special causes. However, in the case of the law of inertia we agreed to dispense with a cause for the displacements (required under a more naïve conception) and were satisfied with having just a cause for the accelerations. I therefore foresee the following development: At such time when physics, and more generally natural science based on observations, shall have completely assimilated the methods and arguments of statistical theory and shall have recognized them as essential tools, the feeling will disappear that these methods and theories contradict any logical need, any 'necessity of thought', or that they leave some philosophical requirement unfulfilled. In other words, the principle of causality is subject to change and it will adjust itself to the requirements of physics.

NEW QUANTUM STATISTICS

Thus there seems to be no serious danger of a permanent check to the development of statistical theory due to arguments of a philosophical nature, or rather due to the adaptation to established routines of thought. It cannot be denied, on the other hand, that the recent development of physical theory has met with difficulties which, in the opinion of a number of contemporary physicists, have not yet been overcome. In the last paragraphs of this book, I will deal briefly with the problems which arise out of the new quantum statistics created by de Broglie,[22] Schroedinger, Heisenberg, and Born.

Let us first recall what was said on the subject of the theory of errors at the end of the last lecture. Each physical measurement is a repetitive event, and can be considered as an element in a collective. A collective is fully described by its distribution, i.e., by the probabilities of all possible values of the attribute, in this case the possible results of the measurement under consideration. As we have seen

211

before, a distribution may be partly characterized by means of two statistical functions, the mean and the variance, which, if the distribution is known, can easily be calculated. The mean value is obtained by multiplying each of the different attributes by its respective probability and adding the products. For an ordinary die the mean value is therefore $1/6 (1 + 2 + 3 + 4 + 5 + 6) = 3.5$. The variance is calculated by subtracting the mean from each value of the attribute, squaring these differences, multiplying each square by the respective probability and summing the products. In the case of a true die the variance equals:

$$1/6(1 - 3.5)^2 + 1/6(2 - 3.5)^2 + 1/6(3 - 3.5)^2 + 1/6(4 - 3.5)^2 \\ + 1/6(5 - 3.5)^2 + 1/6(6 - 3.5)^2 = 1/6 \times 17.5 = 2.92.$$

The mean and the variance are unambiguously determined by the distribution, but a distribution is, of course, not determined by its mean and variance: A number of different distributions have the same mean and the same variance.

In dealing with a collective formed by successive measurements of a physical quantity, the mean is often called the 'true value' of this quantity, a term derived from older conceptions of classical physics. The variance is usually considered as a measure of inexactness or lack of precision of the measurements. Further on we shall see to what extent this is justifiable. The variance by definition equals zero in one case only, namely if the probability of one single attribute equals unity, the probabilities of all other attributes being equal to zero. The value of this attribute is then, of course, also the mean value. Consequently, if in a particular case the variance is very small, we can assume that almost the whole probability is concentrated on values that are close to a certain fixed value, so that it is almost certain that deviations from the 'true value' are very small.

We are now interested in the following question: Is it always possible to indicate a method of measurement of any unknown magnitude such that the resulting variance will be arbitrarily small, and ultimately equal to zero? Classical physics considered this possibility as self-evident and the notion of the 'true value' sprang from this conception. In the first period of physical statistics, in the days of the original kinetic theory of gases, the theories of Brownian motion and of radioactive decay, the current conception in this respect was that a fundamental difference had to be assumed between two orders of magnitude, the 'microscopic' and the 'macroscopic' one. Each of these was made the subject of a separate branch of physics, namely, 'microphysics' and 'macrophysics'. The latter is concerned with variables that are defined in connexion with large numbers of molecules, e.g., the length of a rigid rod, the weight of a body, etc. It was considered that magnitudes of this kind could not be measured with absolute exactness just because they always represent a statistical system, such as a mass of molecules. On the other hand, all magnitudes characteristic of an elementary particle, such as co-ordinates and velocities of a single molecule, were supposed to be exactly and unambiguously measurable, i.e., with a variance equal to zero.

ARE EXACT MEASUREMENTS POSSIBLE?

The new quantum statistics teaches, however, that in the case of elementary particles exact measurements are even less possible. More precisely, it asserts the theoretical impossibility of simultaneous exact determinations of the various magnitudes characterizing a given particle. In fact, from this theory there follows a formula connecting the variances of associated measurements, so that whenever one of the variances becomes very small, a certain other variance must increase. We are now going to study these questions in some detail.

What is the origin of our belief that there are methods of measurement by which a quantity may be determined with any desired degree of accuracy? For instance, let us measure the length of a table with a tape measure. Asked for a result in whole centimetres, we can easily obtain an unambiguous answer. In other words, the repetition of the measurement will always give the same result, say 97 cm. The results form a collective with variance practically equal to zero. We are nevertheless not inclined to call this an exact measurement since the vanishing variance is only achieved by means of selecting a comparatively large unit of measurement. The result 97 cm means only that the length of the object is roughly between 96.5 and 97.5 cm, and we cannot even state that these are precise bounds.

More exact methods of length measurement are most certainly available. To fix a so-called base line, the surveyor uses a very elaborate device permitting him to measure lengths to a thousandth of a millimetre. Repeated readings taken with this instrument give a sequence of integer numbers forming a true collective with a variance well exceeding zero. Nevertheless, this method is considered to be more accurate than the preceding one, because, expressing this result in the customary way, we may say that the length we wish to determine lies within perhaps 0.05 mm. of the measured one. If, working with this instrument, we had recorded the reading to one-tenth of a

213

millimetre only, then the variance would probably again have been reduced to zero.

We can imagine a still more improved measuring device which would permit us to determine with vanishing variance readings to a thousandth of a millimetre or still smaller units. This improvement would, however, not be of much use in surveying. The earth's crust is not rigid enough for a distance of, say, 200 m to remain constant to within 0.001 mm. Such measurements, even those made with the highest degree of accuracy obtainable, belong to the domain of macrophysics, a field in which exact measurements were earlier admitted to be impossible, according to the concepts of classical physics.

Let us now consider the case where the object of measurements is microscopic.

POSITION AND VELOCITY OF A MATERIAL PARTICLE

The physicist, W. Heisenberg, one of the founders of quantum mechanics, was the first to investigate what happens when we try to determine more and more exactly the physical variables characterizing the state of a single particle, i.e., its position in space and its velocity, or its position and its momentum.

First, let us try to fix the position of the particle in space. We place it under the microscope, illuminate it, and try to find its co-ordinates. The exactness with which small objects can be located under a microscope depends on the wave length of the source of illumination. The smallest distance which can be observed under the microscope is proportional to the wave length of the light used. If we want to fix the position as exactly as possible, we have to use light of a very short wave length, and consequently, of a very large frequency.

According to the modern concept of light, an illuminated particle is continuously struck by a large number of light quanta. The whole process is of a statistical nature such as Brownian motion or the motion of molecules in a gas. The energy of each light quantum is inversely proportional to its wave length. The impact of the quanta affects the state of motion of the particle, and this effect increases with the increase in the energy of the quantum, that is, with an increase in its frequency or with a decrease in its wave length (this is the so-called Compton effect). We are thus in a dilemma: the increase in accuracy of the measurement of the co-ordinates of the particle requires the use of light with a very short wave length. The

214

shorter the wave length, however, the stronger is the disturbing influence on the measurement of the velocity of the particle. It follows that it is fundamentally impossible to measure at the same time exactly both the position and the velocity of the particle.

The main point at issue here is not, as has often been stated, that the process of measuring influences the state of the object to be measured and thus limits the possible extent of precision. Such interaction also exists in certain instances of marcophysics, e.g., the introduction of an apparatus for measuring the dynamical pressure of a fluid affects the pressure. However, in this and other such cases we know how to apply appropriate corrections. The conditions in micromechanics are fundamentally different: the essential point is the assumed random character of the disturbing light quanta, a phenomenon which cannot be accounted for by a deterministic theory of the type of Newtonian mechanics.

The essential consequence of Heisenberg's considerations can be summarized by saying that the results of *all* measurements form collectives. In the realm of macrophysics the objects of measurement are themselves statistical conglomerates, such as the length of a ruler which is a mass of molecules in motion. The notion of an absolutely exact length measure has therefore obviously no meaning with respect to objects of this kind. In microphysics, where we are concerned with measurements on a single elementary particle, the inexactness is introduced by the statistical character of the light quanta striking the particle during and through the very act of measuring. In both cases we are faced with the indeterministic nature of the problem as soon as we inquire more closely into the concrete conditions of the act of measuring.

HEISENBERG'S UNCERTAINTY PRINCIPLE[23]

Quantum mechanics is considered today to be a purely statistical theory. Its axioms are expressed in terms of differential equations connecting the probabilities for the values of co-ordinates and velocities at a given moment with the corresponding probabilities at another moment. Some physicists still try to interpret these equations in a deterministic way and to 'derive' them from concepts of classical mechanics to which they are doubtlessly related by many formal analogies. Possibly these attempts will meet with a similar fate as did analogous attempts in the case of Maxwell's equations of electrodynamics. For many years, one tried to explain these equations mechanically, by the introduction of concealed masses and complicated mechanisms. Eventually, however, it was agreed to accept these equations as elementary laws needing no mechanical 'derivation'. The situation is more difficult in the case of quantum mechanics, because here the various assumptions are related to certain mechanical systems.

One consequence of the axioms of quantum mechanics has aroused particular interest. This is the above-mentioned relation existing between the distributions of the co-ordinates of a particle on the one hand and that of its impulses (or velocities) on the other, the most important being that the product of the variances of the two variables has a certain fixed value, independent of any other data of the problem. The order of magnitude of this product is that of the square of Planck's universal constant ($h = 6 \times 10^{-27}$ in the usual metrical units). The relation is known as Heisenberg's Uncertainty Principle. The previously discussed example of the observation of a particle under the microscope, which led to the finding that the more exactly we measure the co-ordinates, the less exact the measurements of the velocities become, appears now as a consequence of Heisenberg's principle.

Heisenberg's principle of the constancy of the product of variances is a purely theoretical proposition and is in this sense mathematically precise. In other words, it presumes that each single measurement in the collective consists in an absolutely exact reading of the measuring instrument. If we were able to make an experimental device to measure lengths to 10^{-13} cm and to measure the impulses also to 10^{-13} gcm/sec, the theory provides that the results of repeated measurements of position will be the same each time (and likewise those of velocity), so that there would be practically no variance in either case. This situation would differ only by its orders of magnitude from the one discussed above where the length of a table was measured without variance by the use of a tape divided into units of whole centimetres only.

Some physicists feel that the ground has been cut from under their feet since the Uncertainty Principle was first announced. If no exact measurements are possible, not even in principle, what is the meaning of exact physical theories? In my opinion, these apprehensions are not justified. The results of quantum mechanics or wave mechanics can be used in exactly the same way as the results of classical macrophysics. What do we care about the impossibility of predicting the beginning of an eclipse of the sun to 10^{-12} seconds, if we can predict it to a second? In the end, our feeling of discomfort is nothing but another aspect of the old disparity between purely mathematical concepts with their 'limitless precision' and the realities of the physical world.

What, then, is the ultimate meaning of Heisenberg's Uncertainty relation? We must see in it a great step towards the unification of our physical conception of the world. Until recently, we thought that there existed two different kinds of observations of natural phenomena, observations of a statistical character, whose exactness could not be improved beyond a certain limit, and observations on the molecular scale whose results were of a mathematically exact and deterministic character. We now recognize that no such distinction exists in nature. I do not want to convey the impression that every distinction between extreme regions of physics has now disappeared, and that the mechanics of solar systems and the theory of radioactive disintegration are only two paragraphs of the same chapter. The description of nature is not as simple as that, and cannot be forced into one single scheme. Nevertheless, a certain apparent contrast between two domains of physics has disappeared with the advent of the new concepts of wave mechanics.

CONSEQUENCES FOR OUR PHYSICAL CONCEPT OF THE WORLD

We can only roughly sketch here the consequences of these new concepts for our general scientific outlook. First of all, we have no cause to doubt the usefulness of the deterministic theories in large domains of physics. These theories, built on a solid body of experience, lead to results that are well confirmed by observation. By allowing us to predict future physical events, these physical theories have fundamentally changed the conditions of human life. The main part of modern technology, using this word in its broadest sense, is still based on the predictions of classical mechanics and physics.

It has been known for a long time, at least to those who strive for clear insight into these matters, that consequences drawn from the mathematical propositions of the classical theories cannot be verified with unlimited accuracy, in the mathematical sense. Atomistic theories of the ancient philosophers already pointed in this direction. The wave theory of light strongly suggests the existence of limitations of this kind. The first attempt at a comprehensive interpretation regarding the nature of the limits to the accuracy of measurements was Boltzmann's formulation, in the second half of the nineteenth century, of the kinetic theory of gases as a statistics of molecules. He pointed out that the predictions of classical physics are to be understood

217

in the sense of probability statements of the type of the Laws of Large Numbers, i.e.: 'If n is a large number, it is almost certain that . . .'. Consideration of the values of n involved, (the number of molecules, etc.), shows that under normal conditions these probabilities are so close to unity that the probable predictions become in fact certain. As explained above, at this stage of development, the usual assumption was that the atomic processes themselves, namely the motions of single molecules, are governed by the exact laws of deterministic mechanics. This point of view which is incompatible with our concept of probability has been retained by some physicists until quite recently.

The rise of quantum mechanics has freed us from this dualism which prevented a logically satisfactory formulation of the fundamentals of physics. We know now that besides classical physics, applicable to processes on a large scale, there is a microphysics, namely the theory of quanta or wave mechanics; the differential equations of microphysics, however, merely connect probability distributions. Therefore, the statements made by this theory with respect to the elementary particles have the character of probability propositions. In the world of molecules, 'exact measurements' without variance are possible only under the same restrictions as hold for ordinary bodies: only if we decide to record just those digits that do not change from one measurement to another. The order of magnitude of the unit, which in atomic physics is about 10^{-12} mm, is of practical but not of basic importance.

I have confined myself to questions regarding inorganic matter and have avoided all attempts to carry the investigations into the field of biology. By this voluntary restriction, I do not intend to indicate that I consider an extension of our theory in this direction to be impossible or impermissible. I think, however, that the so-called biological processes are still much more complicated than those forming the subject of physics and chemistry, and that considerable additions have to be made to the physical theories before biological statements of a basic nature can be attempted.

FINAL CONSIDERATIONS

Let us make a final brief survey of the course which we have followed in these chapters. We began by investigating the meaning of the word 'probability' in everyday language and by trying to restrict this meaning in an appropriate way. We found an adequate basis for the definitions and axioms of an exact scientific theory of probability in a well-known class of phenomena: games of dice and similar processes. The notions of the collective, of the limiting value of relative frequency, and of randomness became the starting-point of the new theory of probability. The four fundamental operations, selection, mixing, partition, and combination, were the tools by means of which the theory was developed.

We stated once and for all that the purpose of the theory is only to derive new distributions of probabilities from initial ones. We showed that, in this sense, the theory of probability does not differ from other natural sciences, and we thus gained a stable position from which to judge the epistemologically insufficient foundations of older theories of probability, like that based on the notion of equally likely events. We reviewed the various suggestions for improvements of my original statements. No necessity for essential alterations emerged from this discussion. The classical Laws of Large Numbers and the recent additions to these laws were incorporated into the new theory. The frequency definition of probability has allowed us to interpret these laws as definite propositions concerning sequences of observable phenomena.

The first wide field of applications of the theory of probability which we have discussed was that usually known as *statistics*. This is, first of all, the study of sequences of numbers derived from the observation of certain repetitive events in human life. We have seen, e.g., that Marbe's exhaustive statistics of the sex distribution of infants is in very good agreement with the predictions of the theory of probability. In other cases, such as death statistics, suicide statistics, the statistical data could not be considered directly as collectives; we found, however, ways to reduce them to collectives. We saw that methods based on the theory of probability, such as, e.g., Lexis's theory of dispersion, were useful tools in a rational comprehensive and systematic description of repetitive events; in this sense, the methods provide us with what is usually called an 'explanation' of the phenomena. The theory of errors, which is the statistics of physical measurements, has served as a link with a second fundamental field of application of the calculus of probability, with *statistical physics*.

The problems of statistical physics are of the greatest interest in our time, since they lead to a revolutionary change in our whole conception of the universe. We have seen how Boltzmann took the first daring step in formulating a law of nature in the form of a statistical proposition. The initial stage was uncertain and in a way self-contradictory in that it attempted to derive the statistical behaviour of systems from the deterministic laws of classical mechanics, an attempt which was destined to fail, as E. Mach maintained vigorously. We have then followed the success of purely statistical arguments in the explanation of certain physical phenomena, such as Brownian motion or the scintillations caused by radioactivity. These investigations led us in a natural way to the problem of the meaning of the so-called law of causality and of the general relation between determinism and indeterminism in physics. We recognized how the progress of physics has brought about a gradual abandonment of preconceived ideas that had even been dogmatically formulated in some philosophical systems. The new quantum mechanics and Heisenberg's Uncertainty Principle finally complete the edifice of a statistical conception of nature, showing that strictly exact observations are no more possible in the world of micromechanics than in that of macromechanics. No measurements can be carried out without the intervention of phenomena of a statistical character.

I think that I may have succeeded in demontrating the thesis indicated in the title and in the introduction to this book: Starting from a logically clear concept of *probability*, based on experience, using arguments which are usually called *statistical*, we can discover *truth* in wide domains of human interest.

SUMMARY OF THE SIX LECTURES IN SIXTEEN PROPOSITIONS

1. The statements of the theory of probability cannot be understood correctly if the word 'probability' is used in the meaning of everyday speech; they hold only for a definite, artificially limited rational concept of probability.

2. This rational concept of probability acquires a precise meaning only if the collective to which it applies is defined exactly in every case. A collective is a mass phenomenon or repetitive event that satisfies certain conditions; generally speaking, it consists of a sequence of observations which can be continued indefinitely.

3. The probability of an attribute (a result of observation) within a collective is the limiting value of the relative frequency with which this attribute recurs in the indefinitely prolonged sequence of observations. This limiting value is not affected by any place selection applied to the sequence (principle of randomness or principle of the impossibility of a gambling system).

Occasionally we deal with sequences in which the condition of randomness is not fulfilled; we then call the limiting value of the relative frequency the 'chance' of the attribute under consideration.

4. The purpose of the calculus of probability, strictly speaking, consists exclusively in the calculation of probability distributions in new collectives derived from given distributions in certain initial collectives. The derivation of new collectives can always be reduced to the (repeated) application of one or several of four simple fundamental operations.

5. A probability value, initial or derived, can only be tested by a statistical experiment, i.e., by means of a sufficiently long sequence of observations. There is no a priori knowledge of probabilities; it is likewise impossible to derive probability values by way of some other non-statistical science, such as mechanics.

6. The classical 'definition' of probability is an attempt to reduce the general case to the special case of equally likely events where all the attributes within the collective have equal probabilities. This reduction is often impossible as, e.g., in the case of death statistics; in other cases it may lead to contradictions (Bertrand's paradox). At any rate, it still remains necessary to give a definition of probability for the case of uniform distributions. Without the complement of a

221

frequency definition, probability theory cannot yield results that are applicable to real events.

7. The so-called Laws of Large Numbers contain meaningful statements on the course of a sequence of observations only if we use a frequency definition of probability. Interpreted in this way, they make definite statements, essentially based on the condition of randomness, concerning the arrangement of the results in the observed sequence. On the basis of the classical definition, these laws are purely arithmetical propositions concerning certain combinatorial properties of integral numbers and bear no relation to the actual evolution of phenomena.

8. The task of probability calculus in mathematical statistics consists in investigating whether a given system of statistical data forms a collective, or whether it can be reduced to collectives. Such a reduction provides a condensed, systematic description of the statistical data that we may properly consider an 'explanation' of these data.

9. None of the theories that seemed to contradict the theory of probability (such as Marbe's theory of statistical stabilization, the theory of accumulation, the law of series) has been confirmed by observations.

10. The concept of likelihood introduced by R. A. Fisher, and the methods of testing derived from it do not, if they are correctly applied and interpreted, fall outside of the domain of the theory of probability based on the frequency concept.

11. The theory of errors, which lies on the borderline between general and physical statistics, is based on the assumption that each physical measurement is an element in a collective whose mean value is the so-called 'true' value of the measured quantity. Additional assumptions concerning this collective lead to the various propositions of the theory of errors.

12. Statistical propositions in physics differ fundamentally from deterministic laws: they predict only what is to be expected in the overwhelming majority of cases for a sufficiently long sequence of observations of the same phenomenon (or of the same group of phenomena). As a rule, however, the relative frequency of this most probable result is so close to unity that no practical difference exists between the statistical proposition and the corresponding deterministic one.

13. Successive observations on the evolution in time of a physical system do not directly form a collective. They can, nevertheless, be dealt with satisfactorily within the framework of the rational theory of probability (probability after-effects, Markoff chains).

14. The assumption that a statistical theory in macrophysics is compatible with a deterministic theory in microphysics is contrary to the conception of probability expressed in these lectures.

15. Modern quantum mechanics or wave mechanics appears to be a purely statistical theory; its fundamental equations state relations between probability distributions. The Uncertainty Principle derived in quantum mechanics implies that measurements in microphysics, like those in macrophysics, are elements of a collective; in either case, a vanishing variance of a measurement is merely the consequence of the choice of a sufficiently large unit of measurement.

16. The point of view that statistical theories are merely temporary explanations, in contrast to the final deterministic ones which alone satisfy our desire for causality, is nothing but a prejudice. Such an opinion can be explained historically, but it is bound to disappear with increased understanding.

NOTES AND ADDENDA

AUTO-BIBLIOGRAPHICAL NOTE

A brief popular presentation of the main ideas of this book can be found in the first part of my article 'Marbe's Gleichförmigkeit in der Welt und die Wahrscheinlichkeitsrechung', *Die Naturwissenschaften* (J. Springer, Berlin) Vol. 7, 1919 No. 11, pp. 168–175; No. 12, pp. 186–192; No. 13, pp. 205–209. A mathematical foundation of the theory was first given in my article 'Grundlagen der Wahrscheinlichkeitsrechnung', *Mathemat. Zeitsch.* (J. Springer, Berlin) Vol. 5 (1919), pp. 52–99. A review of the theory in French can be found in *Annales de l'Institut Henri Poincaré*, Paris, Vol. III, 1932, pp. 137–190 (Conferences held at the Institute in November, 1931). A complete textbook of the theory is *Vorlesungen aus dem Gebiete der angewandten Mathematik*, Bd. I, *Wahrscheinlichkeitsrechnung und ihre Anwendung in der Statistik, Fehlertheorie und in der theoretischen Physik*, 1931 (F. Deuticke, Wien und Leipzig, Reprint, M. Rosenberg, New York, 1945, and the more recent publication: *Mathematical Theory of Probability and Statistics*, Harvard University, Graduate School of Engineering. Special Publ. No. 1, 1946, mimeogr. 320 pp. A brief discussion in English of the basic ideas can be found in the article 'On the foundations of probability and statistics', *Ann. Math. Stat.*, Vol. 12 (1941), pp. 191–205, in Italian, in the paper: 'Sul concetto di probabilità fondato sul limite di frequenze relative', *Giorn. Ist. Ital. Attuari*, Vol. 7 (1936), pp. 235–255, and in French in the lecture, 'Sur les fondéments du calcul des probabilitiés', *Théorie des probabilités, Exposés sur ses fondéments et ses applications*, Paris, Gauthier Villars, 1952, pp. 17–29.

The subject of this book, especially of its last part, is also discussed in the following conferences: 'Über die gegenwärtige Krise der Mechanik', *Zeit f. Angewandte Mathem. und Mechan.*, Vol. I, 1921, pp. 425–431, 'Über kausale und statistische Gesetzmässigkeit in der Physik', *Die Naturwissenschaften*, Vol. 18, 1930, pp. 145–153, and *Erkenntnis (Annalen d. Philosophie)*, Vol. 1, 1930, pp. 189–210, and 'Über das naturwissenschaftliche Weltbild der Gegenwart', Rede beim Stiftungsfest der Berliner Universität, Berlin 1930 (reprinted in *Die Naturwissenschaften*, Vol. 18, 1930, pp. 885–893).

The article by Harald Cramér, 'Richard von Mises' work in probability and statistics', *Ann. Math. Stat.*, Vol. 24 (1953), pp. 657–662, contains a complete bibliography of von Mises' contributions to these subjects.

FIRST LECTURE. DEFINITION OF PROBABILITY

1. GEORG CHRISTOPH LICHTENBERG, *Vermischte Schriften*, Erster Teil, II, I (New edition, Göttingen 1853, Vol. I, p. 79).

2. JACOB and WILHELM GRIMM, *Deutsches Wörterbuch*, Leipzig 1922, Vol. 13, p. 994.

3. R. EISLER, *Wörterbuch der philosophischen Begriffe*, 3rd ed., Berlin 1910, Vol. 3, p. 1743.

4. WEBSTER's *International Dictionary*, Second Ed. Unabridged, 1951, Vol. II, p. 970.

5. W. SOMBART, *Der proletarische Sozialismus*, 10th ed., Jena 1924, Vol. 1, p. 4.

6. I. KANT, *Kritik der reinen Vernunft, Methodenlehre*, 1. Hauptstück, I. Abschnitt, 2nd ed., 1787, p. 758. The later parts of my book do not agree with Kant's theory of definitions outlined in the section quoted here.

7. The best information concerning the concept of work in mechanics and the general problem of the formation of concepts in exact science can be found in E. MACH, *Die Mechanik in ihrer Entwicklung, historisch-kritisch dargestellt* (1883), 7th ed., Leipzig 1921, which is also suitable for a non-mathematical reader. The same problem is discussed even more precisely in E. MACH, *Prinzipien der Wärmelehre, historisch-kritisch entwickelt* (1896), 4th ed., Leipzig 1923, pp. 406–432 (Abschnitte: Die Sprache, Der Begriff, Der Substanzbegriff). The point of view represented in this book corresponds essentially to MACH's ideas.

8. Concerning the controversy on the 'true measure of force' see, e.g., MACH, *Mechanik* (cit. above), pp. 247 and 288. The controversy, initiated by Leibniz, continued for 57 years and was only settled in 1743 by D'ALEMBERT's *Traité de dynamique*.

9. GOETHE's article in *Propyläen*, Vol. I, No. 1 (Werke, Ausgabe letzter Hand 12°, Vol. 38, 1830, pp. 143–154) uses the word 'probability' in the sense of 'illusion'. In doing so, he shows a much finer sense of language than the philosophers previously quoted. 'Eine auf dem Theater dargestellte Szene muss nicht *wahr scheinen*, aber einen *Schein von Wahrheit* vermitteln'.

10. A. A. MARKOFF says in his textbook, *Theory of Probability* (German edition by H. Liebmann, Leipzig and Berlin 1912, p. 199): 'We do not agree at all with the academician Bunjakowski (*Foundations of the mathematical theory of probability*, p. 326) who says that a certain class of narratives must remain unconsidered, because it is not permitted to doubt their truth'.

11. PIERRE SIMON (later Marquis de) LAPLACE (1749–1827) published in 1814 his *Essai philosophique des probabilités*, which was reprinted as 'Introduction' in the later editions of his *Théorie analytique des probabilités*, (1st ed. 1812, 2nd ed. 1814, 3rd ed. 1820). The *Essai* represents the point of view of unlimited determinism and is a characteristic expression of the

225

philosophical school of 18th century France. The *Essai* was republished in a convenient form in 1921 in the series *Les Maîtres de la Pensée scientifique*, Paris, 1921.

12. SIMÉON DENIS POISSON (1781–1840), published in 1837 a mathematical textbook of the theory of probability under the title *Recherches sur la probabilité des jugements en matière criminelle et en matière civile*. The subject mentioned in the title is only treated in the fifth chapter of the book, which is one of the most important works in the history of the mathematical theory of probability.

13. A mathematical definition of the concept of limiting value may be given in the following form: We say, that an infinite sequence of numbers $a_1, a_2, \ldots$ lying between 0 and 1 approaches a limiting value if, no matter how large k may be, beginning with a certain number, a_N (where N depends on k) all those following have the same k first figures after the decimal point.

14. *Deutsche Sterblichkeitstafeln aus den Erfahrungen von 23 Lebensversicherungsgesellschaften*, Berlin 1883. Short remarks about this and other similar tables can be found in E. CZUBER, *Wahrscheinlichkeitsrechnung*, 3rd ed., Leipzig and Berlin 1921, Vol. 2, p. 140.

15. J. V. KRIES, *Die Prinzipien der Wahrscheinlichkeitsrechnung, eine logische Untersuchung* (1886), second reprint, Tübingen 1927, p. 130.

16. LAPLACE, *Essai* (see note 11), p. 15 of the 1921 French edition.

17. POISSON'S textbook has been cited in footnote 12.

18. LAPLACE, see note 11.

19. JOHN VENN, *The logic of chance*, London and Cambridge 1866.

20. TH. FECHNER, *Kollektivmasslehre*, edited by A. F. Lipps, Leipzig 1897.

21. H. BRUNS, *Wahrscheinlichkeitsrechnung und Kollektivmasslehre*, Leipzig and Berlin, 1906.

22. G. HELM, 'Die Wahrscheinlichkeitslehre als Theorie der Kollektivbegriffe', *Annalen der Naturphilos.*, Vol. I, 1902, pp. 364–381.

SECOND LECTURE: THE ELEMENTS OF THE THEORY OF PROBABILITY

1. In the article of E. CZUBER in *Enzyklop. der mathem. Wissensch.*, Vol. I (2nd part), p. 736, we read about the subjectivists: 'According to the first principle (i.e., that of insufficient reason) the statement of equal likelihood is founded on the absolute lack of knowledge concerning the conditions of existence or realization of the single cases. . . .'

2. A reader acquainted with mathematics will note that the probability density of an attribute is the derivative of the probability with respect to the variables determining the attribute. Let us denote by x the possible numerical values of a physical magnitude which we are going to measure, and by $w(x)$ the probability density; $w(x)dx$ is then the probability of the

experimental value falling into the interval x to $x + dx$. The density function must fulfil the condition

$$\int_{-\infty}^{\infty} w(x)dx = 1.$$

3. The mixing rule for continuous distributions can be formulated as follows: The probability of a magnitude with a probability density function $w(x)$ falling into the interval between $x = a$ and $x = b$ equals

$$\int_{a}^{b} w(x)dx.$$

4. TH. BAYES's famous memoir was first published by R. Price after the author's death, under the title 'An essay towards solving a problem in the doctrine of chances', London, *Philos. Trans.* 53 (1763), pp. 376–398 and 54 (1764), pp. 298–310. A new edition was published by W. E. DEMING, *Facsimiles of two papers by Bayes*, Washington (1940), the Department of Agriculture.

5. Our knowledge of this problem is due to the preservation of the letters exchanged by FERMAT and PASCAL. They are reported, together with many other historical facts of the pre-Laplacian period, in J. TODHUNTER, *A history of the mathematical theory of probability, From the time of Pascal to that of Laplace*, reprint, New York, G. E. Stechert, 1931.

THIRD LECTURE: CRITICAL DISCUSSION OF THE FOUNDATIONS OF PROBABILITY

1. In reference to this, it may be of interest to consider the mathematical part (3) of WEBSTER's definition (see note 4, Lect. I): 'In the doctrine of chance, the likelihood of the occurrence of any particular form of an event, estimated as the ratio of the number of ways in which that event might occur to the whole number of ways in which the event might occur in any form (all such elementary forms being assumed as equally probable); the limit of the ratio of the frequency of that form of the event to the entire frequency of the event in all forms as the number of trials is increased indefinitely. Thus, as an unweighted die thrown up may fall equally well with any of its six faces up, there are 6 ways of happening; the ace can turn up in only 1 way; the chance of the ace is 1 out of 6 (1/6).' We have here, at first, a rather awkward formulation of Laplace's definition; this is followed by a kind of frequency definition which, however, is not clearly distinguished from the former since by means of the word 'thus' Webster returns to the unbiased die and the equally likely cases.

2. HENRI POINCARÉ, *Calcul des probabilités*, 1. éd., Paris 1912, p. 24. A review of the literature on the foundations of the theory of probability (up

to about 1916) is given by E. CZUBER in his book, *Die philosophischen Grundlagen der Wahrscheinlichkeitsrechnung*, Leipzig and Berlin 1923. For the early history of the theory see also J. TODHUNTER (cit. note 5, Lect. II).

3. JACOB BERNOULLI, *Ars conjectandi*, Basel 1713, e.g., p. 6, propositio II: 'si *a*, *b* vel *c* expectem, quorum unumquodque pari facilitate mihi obtingere possit, expectatione mea aestimanda est $(a + b + c)/3$.'

4. A. MEINONG, *Über Möglichkeit und Wahrscheinlichkeit, Beiträge zur Gegenstands—und Erkenntnistheorie*, Leipzig 1915.

5. In LAPLACE'S *Théorie analytique* (note 11, Lect. I), the title of Ch. 7 (1st ed., pp. 402–407), is as follows: 'De l'influence des inégalités inconnues qui peuvent exister entre des chances que l'on suppose parfaitement égales.' There he says: '. . . $(1 + \alpha)/2$ soit la possibilité d'amener pile . . .'; we see that he uses the word 'possibilité' in place of 'probabilité', because the original definition of 'probabilité' is no longer suitable. Further on in the text, however, the rules derived previously for 'probabilité' are applied without scruple to 'possibilité'.

6. The principal scientific centre for the study of parapsychology in the United States is probably the Parapsychological Laboratory at Duke University, N.C. whose findings are published in the *Journal of Parapsychology*, founded in 1937. It is interesting to us that by means of certain statistical experiments, among which a type of card calling experiment is prominent, the existence of so-called extra-sensorial perception, ESP, is investigated with the claim that positive results have been obtained. In a lecture at the AAAS meeting (Dec. 30, 1949, New York City), von Mises attempted to show that so far this aim has hardly been reached and pointed out certain ways to modify the setup of experiments and observations in order to obtain more decisive results.

7. J. M. KEYNES, *Treatise on probability*, London 1921. As a characteristic representative of the subjectivists, we may consider C. STUMPF, 'Über den Begriff der mathematischen Wahrscheinlichkeit', *Sitz.-Berichte der Bayr. Ak. d. Wiss., philos.-hist. Klasse* 1892, p. 41. We find here the following detailed reference to the part played by lack of knowledge: 'Gleich möglich sind Fälle, in bezug auf welche wir uns in gleicher Unwissenheit befinden. Und da die Unwissenheit nur dann ihrem Mass nach gleichgesetzt werden kann, wenn wir absolut nichts darüber wissen, welcher von den unterscheidenden Fällen eintreten wird, so können wir noch bestimmter diese Erklärung dafür einsetzen.'

8. Bertrand's Paradox is discussed in most of the textbooks on probability. See, for instance, J. V. USPENSKY, *Introduction to mathematical probability*, New York and London 1937, p. 251, where the solution of this apparent paradox is given in our sense, or my textbook (cit. autobibliogr. note), pp. 78–83.

9. For BUFFON's own discussion see *Histoire de l'Académie des Sciences*, Paris 1733, pp. 43–45, also his book *Essai d'arithmétique morale*, Paris 1775.

10. In speaking of the 'older generation', I have in mind especially a

group of Italian mathematicians who in several papers which appeared in 1916–17 (i.e., before the publication of my investigations) claimed to have proved the 'inadmissibility' of the assumption of limiting values of relative frequencies. See, for instance, F. P. CANTELLI, *Annal. de l'Institut Henri Poincaré*, Vol. 5, 1935, pp. 1–50. What is actually proved in this paper is the contradiction which arises if one assumes the existence of the limiting value of relative frequencies while using the term probability in calculations and applications without postulating its identity with limiting frequency. In answer, see: R. de Misès, Sul concetto di probabilità fondato sul limite di frequenze relative, *Giorn. dell'Istituto Ital. degli Attuari* 1936, pp. 235–255.

11. FRECHET et HALBWACHS, *Le calcul des probabilités à la portée de tous*, Paris 1924.

12. J. L. COOLIDGE, *An introduction to mathematical probability*, Oxford 1925.

13. HARALD CRAMÉR, *Mathematical methods of statistics*, Stockholm 1945 and Princeton 1946. See in particular p. 150ff.

14. HANS BLUME, *Zur axiomatischen Grundlegung der Wahrscheinlichkeitsrechnung*, 1934 (Dissert. Munster). Also two papers in *Zeitsch. f. Physik*, Vol. 92 (1934), pp. 232–252, and Vol. 94 (1935), pp. 192–203.

15. A. KOLMOGOROFF's criticism appeared in *Zentralblatt f. Mathem.*, Vol. 10 (1935), p. 172.

16. Subsequent to the author's last revision of this text, A. H. COPELAND published the following paper on this subject: A finite frequency theory of probability. *Studies in Math. and Mech.* Presented to Richard von Mises, New York, 1954.

17. THEODOR FECHNER, see note 20, Lect. I.

18. CARL G. HEMPEL, '*Erkenntnis*' (*Annalen der Philosophie*), Vol. 5 (1935), pp. 228–260.

19. This objection forms the main content of the criticism of CANTELLI and other Italians referred to in note 10, this Lect.

20. A. H. COPELAND, Independent event histories, *Am. J. of Math.*, 51 (1929), pp. 612–618; The theory of probability from the point of view of admissible numbers, *Ann. of Math. Stat.*, 3 (1932), pp. 143–156; Admissible numbers in the theory of geometrical probability, *Am. J. of Math.*, 53 (1931), pp. 153–162.

21. R. V. MISES, Über Zahlenfolgen die ein kollektiv-ähnliches Verhalten zeigen, *Math. Ann.*, Vol. 108 (1933), pp. 757–772. On the subject of Bernoulli-sequences cf. also H. REICHENBACH, *Wahrscheinlichkeitslehre*, Leiden 1935. Engl. ed., *The theory of probability*, Berkeley and Los Angeles 1945.

22. K. DÖRGE, Eine Axiomatisierung der von Misesschen Wahrscheinlichkeitstheorie, *Jahresb. d. deutschen Mathematiker-Vereinigung*, 43 (1934), pp. 39–47. Also, by the same author, *Mathem. Zeitsch.*, 32 (1930), pp. 232–258 and 40 (1935), pp. 161–193.

23. A. H. COPELAND, besides the papers indicated in note 20 above, see

also, Point set theory applied to the random selection of the digits of an admissible number, *Amer. J. of Mathem.*, 58 (1936), pp. 181–192.

24. A. WALD, Über die Wiederspruchsfreiheit des Kollektivbegriffes, *Ergebnisse eines mathem.* Kolloquiums, Wien No. 8, pp. 38–72.

25. W. FELLER, Über die Existenz sogenannter Kollektive, *Fundamentæ Mathematicæ*, 32 (1939), pp. 87–96.

26. ED. V. HARTMANN, *Philosophie des Unbewussten*, 11th ed., Leipzig 1904, Vol. 1, pp. 36–47.

27. J. M. KEYNES, *Treatise on Probability*, London 1921.

28. HAROLD JEFFREYS, *Scientific Inference*, Cambridge 1931; esp. pp. 8–35.

29. G. PÓLYA, Heuristic reasoning and the theory of probability, *Amer. Mathem. Monthly*, 48 (1941), pp. 450–465. More recent works by the same author on this subject are: On patterns of plausible inference, *Courant Anniversary Volume*, 1948, pp. 277–288. Preliminary remarks on a logic of plausible inference, *Dialectica*, Vol. 3 (1949), pp. 28–35. On plausible reasoning, *Proc. Intern. Congr. of Math.*, 1950, Vol. I, pp. 739–747. *Mathematics and Plausible Reasoning*, Vols. I and II, Princeton University Press 1954; note esp. Chs. XIV and XV.

30. R. CARNAP, *Logical foundations of probability*, University of Chicago Press, Vol. 1, 1950.

31. C. G. HEMPEL and P. OPPENHEIM, A definition of 'degree of confirmation', *Philosophy of Science*, 12 (1945), pp. 98–115.

32. In a paper, 'Über die J. v. Neumann'sche Theorie der Spiele', *Mathem. Nachrichten*, 9 (1953), von Mises comments: 'In the detailed presentation given in the book of v. Neumann and Morgenstern (*Theory of games and economic behaviour*, Princeton 1944) where every arithmetic or geometric concept is analyzed down to its last element, the words "probability" and "expected value" are used without any definition.'

33. A. KOLMOGOROFF, Grundbegriffe der Wahrscheinlichkeitsrechnung, *Ergebnisse der Mathematik und ihrer Grenzgebiete*, Vol. 2, No. 3, Berlin 1933. Engl. transl. *Foundations of the Theory of Probability*, New York 1950.

34. Here some ten lines of the text were replaced by the editor by other material taken from the author's writings which, in her opinion, render more clearly v. Mises' point of view in his last years, see particularly: *Sur les fondéments du calcul des probabilités . . .*, cit. in auto-bibliogr. note.

35. Review by DOETSCH, *Jahresber. d. deutsch. Mathem. Verein.*, Vol. 45 (1935), p. 153.

36. From E. TORNIER's writings we may cite here: Wahrscheinlichkeitsrechnung und Zahlentheorie, *Journal f. die reine und angewandte Mathem.*, Bd. 160 (1929) pp. 177–195, Die Axiome der Wahrscheinlichkeitsrechnung, *Acta Mathematica*, 60 (1939), pp. 239–280; *Wahrscheinlichkeitsrechnung und Integrationstheorie*, Leipzig 1936; *Theorie der Versuchsvorschriften der Wahrscheinlichkeitsrechnung*, by ERHARD TORNIER and HANS DOMIZLAFF, Stuttgart 1952.

37. J. L. DOOB, Note on probability, *Annals of Mathematics*, 37 (1936), pp. 363–367.

FOURTH LECTURE: THE LAWS OF LARGE NUMBERS

1. The essential content of this lecture, up to about p. 112, was first published by the author in *Die Naturwissenschaften*, Vol. 15 (1927), pp. 479–502.

2. POISSON, see note 12, Lect. I. The quotation from Poisson's introduction is translated from p. 7 of the original.

3. JACOB BERNOULLI, *Ars conjectandi*, Basel 1713. The law in question is found in Part IV, Ch. 5, p. 236.

4. P. L. TSCHEBYSCHEFF's general proposition was first published in 1867 in Russian; it appeared later in *Jour. de Liouville*, sér. II, Vol. 12 (1867). The derivation is elementary throughout.

5. The *deus ex machina* is introduced quite openly, e.g., by H. WEYL, Philosophie der Mathematik und Naturwissenschaft, *Handbuch der Philosophie*, Munich and Berlin 1927, p. 151.

6. Mathematical problems of a similar type are discussed by G. PÓLYA and G. SZEGOE, *Aufgaben und Lehrsätze aus der Analysis*, Vol. I, Berlin 1925, pp. 72 and 238. The structure of the table of logarithms (referred to on p. 111) is also explained in this book. Further examples and explanations of other sequences of numbers pertinent to this problem can be found in my paper: Über Zahlenreihen die ein kollektivähnliches Verhalten zeigen, see auto-bibliogr. note.

7. BAYES's original paper was quoted above, see note 4, Lect. II. It does not contain 'Bayes's Theorem' as such; the name is justified by the fact that the problem was suggested by Bayes's. The solution as well as the name are due to Laplace.

8. This stronger form of the Law of Large Numbers was first given by F. P. CANTELLI (1917). It was found independently by G. PÓLYA in 1921. On this subject see also, A. KHINTCHINE, Über einen Satz der Wahrscheinlichkeitsrechnung, *Fundamenta Mathematica*, 6 (1929), pp. 9–20. The explanation given by us follows the line indicated by A. KOLMOGOROFF, Das Gesetz des iterierten Logarithmus, *Mathem. Ann.*, 101 (1929), pp. 126–135. See also the presentation in WM. FELLER, *An introduction to probability theory and its applications*, Vol. 1, New York and London 1950.

9. This extension of the Law of Large Numbers was first indicated in my 'Vorlesungen', cit. above, pp. 192–197. The general concept of statistical functions was first introduced in the article 'Deux nouveaux théorèmes de limite dans le calcul des probabilités', *Rev. de la Faculté des Sciences d'Istanbul*, Vol. 1 (1935), pp. 61–80. A detailed derivation is given in 'Die Gesetze der grossen Zahlen für statistische Funktionen', *Monatshefte für Mathem. u. Physik*, Vol. 43 (1936), pp. 105–128. See also: R. V. MISES, On the asymptotic distribution of differentiable statistical functions, *Annals of Mathem. Statistics*, 18 (1947), pp. 309–348.

PROBABILITY, STATISTICS AND TRUTH

FIFTH LECTURE: APPLICATIONS IN STATISTICS AND THE THEORY
OF ERRORS

1. KARL MARBE, *Die Gleichförmigkeit in der Welt, Untersuchungen zur Philosophie und positiven Wissenschaft*, Munich 1916, and by the same author, *Mathematische Bemerkungen zu meinem Buch 'Die Gleichförmigkeit in der Welt'*, ibid. 1916. The quotation is from Vol. 1, p. 375.

2. KARL MARBE, *Grundfragen der angewandten Wahrscheinlichkeitsrechnung und theoretischen Statistik*, Munich and Berlin, 1934. I have published a review of this book in *Die Naturwissenschaften*, Vol. 22 (1934), pp. 741–743.

3. Compare my article (cited in the auto-bibliogr. note) in *Die Naturwissenschaften*, 1919, and the mathematical investigation of the problem in 'Zur Theorie der Iterationen', *Zeitsch. f. angew. Mathem. u. Mechan.*, Vol. 1 (1921), pp. 298–307. It is shown there that the probability of the occurrence of x runs of length m in n single observations (of two attributes) is given by the formula

$$p = \frac{e^{-a} \times a^x}{x!}$$

where $a = n \times 2^{m-1}$.

4. O. STERZINGER, *Zur Logik und Naturphilosophie der Wahrscheinlichkeitslehre*, Leipzig 1911.

5. PAUL KAMMERER, *Das Gesetz der Serie, eine Lehre von den Wiederholungen im Lebens—und im Weltgeschehen*, Stuttgart and Berlin 1919.

6. F. EGGENBERGER and G. PÓLYA, 'Über die Statistik verketteter Vorgänge', *Zeitsch. f. angew. Mathem. u. Mechan.*, Vol. 3 (1923), pp. 279–289.

7. W. LEXIS, *Zur Theorie der Massenerscheinungen in der menschlichen Gesellschaft*, Freiburg, i. B. 1877. A description of this theory is given in many textbooks on probability calculus. e.g., in the well-known book of H. L. RIETZ, *Mathematical Statistics*, Chicago 1927.

8. The figures are from *Oesterreichische Statistik*, Vol. 88 (1911), No. 3, pp. 20 and 120.

9. The algebraic theorem is the so-called Schwarz inequality: If $p_1, p_2,$ . . . p_z are the single probabilities, and p is their mean value, then:

$$p_1(1 - p_1) + p_2(1 - p_2) + \ldots p_z(1 - p_z) \leq zp(1 - p).$$

10. From *Statistisches Jahrbuch für das Deutsche Reich*, Vol. 43 (1923), p. 35.

11. R. A. FISHER, 'On the mathematical foundations of theoretical statistics', *Philos. Trans. of the Royal Society*, London, A 222 (1921), p. 309. Cf. also: 'The concept of inverse probability and the use of likelihood', *Proc. Cambridge Philos. Soc.*, 28 (1932), p. 257.

12. Cf. e.g., O. HELMER and P. OPPENHEIM, A syntactical definition of probability and degree of confirmation, *J. of Symbolic Logic*, 10 (1945), pp. 25–60.

232

13. This example is taken from P. R. RIDER, Small sample theory, *Ann. of Mathem.* II, 31 (1930), p. 577.

14. A. QUETELET, *Physique sociale ou essai sur le développement des facultés de l'homme*, Brussels, Paris, St. Petersburg, 1869.

15. For an English translation of Mendel's work see: G. MENDEL, *Experiments on Plant Hybridization*, Harvard U.P., Cambridge 1946. For the numerical examples see E. CZUBER, *Statistische Forschungsmethoden*, Vienna 1921, p. 184.

16. On the application of statistics to technology see: T. C. FRY, *Probability and its engineering uses*, New York 1928; A. HALD, *Statistical theory with engineering applications*, New York 1952, which deals, among others, with the statistical problems of telephone communications. Regarding quality control, see W. A. SHEWHART, *Economic control of quality of manufactured product*, New York 1931, and, by the same author, *Statistical method from the viewpoint of quality control*, The Graduate School Dep't. of Agriculture, Washington, D.C. 1939. Also the comprehensive work of E. L. GRANT, *Statistical quality control*, New York 1952.

17. Problems of this type, taken from various fields of applications, have met with considerable interest in recent years, they are grouped under the name of Operational Research or Operations Research. See, e.g., *M.I.T. Course on Operational Research*, Technology Press 1952; MORSE and KIMBALL, *Methods of Operational Research*, 1950; *J. Operations Research Soc. of America*, 1953. A very interesting idea has found wide application in the field of operations research and also to a certain extent in problems of pure mathematics, the so-called Monte Carlo method. 'Random numbers' are used in order to create an artificial world of experience attempting to find approximate answers to given problems through statistical experimentations. This is made practicable by the use of large scale computing machines. See, e.g., D. M. MCCRACKEN, The Monte Carlo Method, *Scientific Am.*, May 1955, p. 90ff; W. E. MILNE, *Numerical Solution of Differential Equations*, J. Wiley, 1953, Apps. B and C; JOHN H. CURTISS, *Sampling Methods applied to differential and difference equations*, Report of the November 1949 Seminar on Scientific Computation, I.B.M. Corp.; 'Monte Carlo Method', Proc. of a symposium held June 29, 30, and July 1, 1949 in Los Angeles, Cal., printed in *Nat'l. Bureau of Standards, Appl. Math. Series* 12 (1951).

18. This deduction is to be found in E. BLEULER, *Das autistisch-undisziplinierte Denken in der Medizin und seine Überwindung*, Berlin 1921, pp. 132–145, which is on the whole a brilliant book. The point of view of the mathematician G. PÓLYA is given *ibid.*, pp. 145–148.

19. Statistical measures in current use are described in G. U. YULE, M. G. KENDALL, *An introduction to the theory of statistics*, Twelfth ed., London 1940. A non-mathematician may consult F. ZIZEK, *Die statistischen Mittelwerte*, Leipzig 1908, Engl. transl., *Statistical averages*, New York 1913.

20. For GINI's 'measure of disparity' see his discussion in *Atti del R.*

Instituto Ven. di scienze, lettere ed arti, t. 73, II, pp. 1203–1248, and the discussion by L. V. BORTKIEWICZ in *XIXe Session de l'Institut International de Statistique*, Tokio 1930, The Hague 1930, Vol. II, pp. 1–108.

21. Pearson's types of distributions are described, e.g., by W. PALIN ELDERTON in *Frequency curves and correlation*, London 1938.

22. H. BRUNS, *Wahrscheinlichkeitsrechnung und Kollektivmasslehre*, Leipzig and Berlin 1906; C. V. L. CHARLIER, *Vorlesungen über die Grundzüge der mathematischen Statistik*, Lund 1920. The rational method of description mentioned in the text consists in expanding the distribution function into an infinite sequence whose coefficients are the 'characteristic numbers'. Bruns's development is based on Gauss's e^{-x^2} function, Charliers' on Poisson's function $a^x e^{-x}/x!$. The subsequent terms are the differential quotients of the first one or the difference quotients respectively. The mathematical theory of Bruns's sequence has been given in my paper in *Jahresber. d. deutsch. Mathem. Verein.*, Vol. 21 (1912), pp. 9–20; that of Charlier's sequence in H. POLLACZEK-GEIRINGER's paper in *Skandinavsk Aktuarietidskrift*, 1928, pp. 98–111. Cf. also the latter's article 'Die Statistik seltener Ereignisse' in *Die Naturwissenschaften*, Vol. 16 (1928), pp. 815–820.

23. C. F. GAUSS's fundamental papers (1821 and 1826) have been published in German by A. Borsch and P. Simon, Berlin 1887 under the title *Abhandlungen zur Methode der kleinsten Quadrate*.

24. LAPLACE's theorem, mentioned in this paragraph, is one of the foremost subjects of mathematical investigations in the field of statistics and theory of probability. Cf. my article 'Fundamentalsätze der Wahrscheinlichkeitsrechnung', *Mathem. Zeitsch.*, Vol. 4 (1919), pp. 1–96. More recent developments are described in A. KHINTCHINE, *Asymptotische Gesetze der Wahrscheinlichkeitsrechnung*, Ergebnisse der Mathematik und ihrer Grenzgebiete, Vol. 2, No. 4, Berlin 1933. In English, New York, Chelsea Publishing Co., 1948. See also the up-to-date presentation by B. V. GNEDENKO, A. N. KOLMOGOROV, *Limit distributions for sums of independent random variables* (transl. from the Russian by K. L. Chung), Cambridge 1954.

25. Most of the investigations of K. PEARSON and his school have been published in *Biometrica, a journal for the study of biological problems*, founded in 1902. The most important work of K. PEARSON is his *Grammar of Science*, London 1900.

SIXTH LECTURE: STATISTICAL PROBLEMS IN PHYSICS

1. BOLTZMANN's theory is presented in his *Vorlesungen über Gastheorie*, 2 vols, 2nd impression, Leipzig 1910. The main ideas are discussed in all textbooks of theoretical physics.

2. M. V. SMOLUCHOWSKI, 'Über den Begriff des Zufalls und den Ursprung der Wahrscheinlichkeitsgesetze in der Physik', *Die Naturwissenschaften*, Vol. 6, 1921, pp. 253–263.

3. LAPLACE's 'demon' is introduced in his *Essai philosophique* (note 11, Lect. I).

4. H. TH. BUCKLE, *History of Civilization in England*, 2nd ed., London 1858. In this book we find information about the beginnings of statistics.

5. M. V. SMOLUCHOWSKI, see note 2, above.

6. H. POINCARÉ, *Science et méthode*, éd. définitive, Paris, p. 71, Engl. transl., *Science and method*, New York 1908.

7. The use of the terms 'million', 'billion', etc. is not unambiguous. In German (the original language of the present book), 1,000 millions, i.e., 10^9 is called a milliard; in American English the same number is called a billion, whereas in British English, as well as in German, a billion means 10^{12} or a million millions. The use of the word trillion is similarly confused. Our text uses the following terms:

$$10^6 = 1,000,000 = 1 \text{ million}$$
$$10^{12} = 10^6 \times 10^6 = 1 \text{ billion}$$
$$10^{18} = 10^6 \times 10^{12} = 1 \text{ trillion.}$$

The number n of molecules in 1 cm^3 at pressure p of one atmosphere, temperature $T = 273°$ is (Millikan) $n = 2.705 \times 10^{19}$. This is approximated in our text as $3 \times 10^{19} = 30,000 \times 10^{15}$. But $30,000 \times 10^{15}$ per cm^3 equals $30,000 \times 10^{12}$ or 30,000 billions per mm^3.

8. For an introduction to the older atomic theory, with many numerical data, we mention J. PERRIN, *Les atomes*, Paris 1914. Engl. transl. *Atoms*, London 1923.

9. ERNST MACH, *Die Leitgedanken meiner naturwissenschaftlichen Erkenntnislehre und ihre Aufnahme durch die Zeitgenossen*, Leipzig 1919, p. 9; see also *Wärmelehre*, 4th ed., p. 364.

10. The fundamental investigations of M. V. SMOLUCHOWSKI are published in *Sitzungsber. der Wien. Akad. d. Wiss.*, math.-naturw. Kl., Abt. IIa, Vol. 123 (1914), pp. 2381–2405. and Vol. 124 (1915), pp. 339–368.

11. The clarification of the basic problem of the evolution of this and similar phenomena in time was given by the author in the paper: 'Ausschaltung der Ergodenhypothese in der physikalischen Statistik', *Physik. Zeitsch.*, 21 (1920), pp. 225–232 and pp. 256–262.

12. Regarding the important role of MARKOFF's chains for physical statistics, see my *Vorlesungen . . .*, p. 16 (cit. auto-bibliogr. note).

13. SVEDBERG's experiments are described in T. SVEDBERG, *Existenz der Moleküle*, Leipzig 1912, p. 148. See also R. FÜRTH, *Schwankungserscheinungen in der Physik*, Braunschweig 1920.

14. In this connexion, see L. V. BORTKIEWITZ, *Die radioaktive Strahlung als Gegenstand Wahrscheinlic keitstheoretischer Untersuchungen*, Berlin 1913.

15. E. MARSDEN and T. BARRAT, *Proc. Phys. Soc.*, London, Vol. 23 (1911), pp. 367–373.

16. Information about recent developments in the theory of gases, the electronic theory of metals, and problems of the quantum theory which

will be discussed later in this book can be found in F. K. RICHTMYER and E. H. KENNARD, *Introduction to modern physics*, New York 1947.

17. The fundamental investigations by M. PLANCK, 'Theorie des Gesetzes der Energieverteilung im Normalspektrum' appeared in *Verkandl. d. Deutsch. Phys. Gesellsch.*, Vol. 2 (1900), pp. 237–245, and in *Annal. der Physik*, Vol. 4 (1901), pp. 553–563.

18. PHILLIP FRANK, *Das Kausalgesetz und seine Grenzen* (*Schriften zur wiss. Weltauff.*), Vol. 6, Wien 1932.

19. In connexion with the following discussion, compare my lecture: 'Über die gegenwärtige Krise in der Mechanik', 1921, (cit. auto-bibliogr. note). There I demonstrated for the first time the incompatibility of a deterministic theory 'on a small scale' with a statistical theory 'on a large scale'.

20. See my lecture, *Naturwissenschaft und Technik der Gegenwart*, Leipzig and Berlin, 1922; also *Zeitschr. der Verein. deutsch. Ingen.*, Vol. 64 (1920), pp. 687–690 and 717–719.

21. This section is taken from my lecture: *Über kausale und statistische Gesetzmässigkeit*, etc., 1930 (cit. auto-bibliogr. note).

22. Original works: L. DE BROGLIE, *Introduction à l'étude de la mécanique ondulatoire*, Paris, 1930. (Engl. transl., *An Introduction to the study of wave mechanics*, London 1930.) W. HEISENBERG, *Die physikalischen Prinzipien der Quantentheorie*, Leipzig 1930. (Engl. transl., *The Physical Principles of the Quantum Theory*, New York 1949.) E. SCHRÖDINGER, *Abhandlunger zur Wellenmechanik*, Leipzig 1927. (Engl. transl., *Four Lectures on Wave Mechanics*, London 1928.) A popular review is found in RICHTMYER and KENNARD, see note 16, above.

23. HEISENBERG's Uncertainty Principle is interpreted in different ways by physicists: cf. for instance, M. V. LAUE, *Die Naturwissenschaften*, Vol. 22 (1934), p. 439, and my answer, *ibid.*, p. 822.

SUBJECT INDEX*

* References to Pages 224–236 are found in the Notes.

NAME INDEX

PROBABILITY, STATISTICS AND TRUTH

A CATALOG OF SELECTED
DOVER BOOKS
IN ALL FIELDS OF INTEREST

A CATALOG OF SELECTED DOVER
BOOKS IN ALL FIELDS OF INTEREST

CONCERNING THE SPIRITUAL IN ART, Wassily Kandinsky. Pioneering work by father of abstract art. Thoughts on color theory, nature of art. Analysis of earlier masters. 12 illustrations. 80pp. of text. 5⅜ × 8½. 23411-8 Pa. $3.95

ANIMALS: 1,419 Copyright-Free Illustrations of Mammals, Birds, Fish, Insects, etc., Jim Harter (ed.). Clear wood engravings present, in extremely lifelike poses, over 1,000 species of animals. One of the most extensive pictorial sourcebooks of its kind. Captions. Index. 284pp. 9 × 12. 23766-4 Pa. $12.95

CELTIC ART: The Methods of Construction, George Bain. Simple geometric techniques for making Celtic interlacements, spirals, Kells-type initials, animals, humans, etc. Over 500 illustrations. 160pp. 9 × 12. (USO) 22923-8 Pa. $9.95

AN ATLAS OF ANATOMY FOR ARTISTS, Fritz Schider. Most thorough reference work on art anatomy in the world. Hundreds of illustrations, including selections from works by Vesalius, Leonardo, Goya, Ingres, Michelangelo, others. 593 illustrations. 192pp. 7⅛ × 10¼. 20241-0 Pa. $9.95

CELTIC HAND STROKE-BY-STROKE (Irish Half-Uncial from "The Book of Kells"): An Arthur Baker Calligraphy Manual, Arthur Baker. Complete guide to creating each letter of the alphabet in distinctive Celtic manner. Covers hand position, strokes, pens, inks, paper, more. Illustrated. 48pp. 8¼ × 11. 24336-2 Pa. $3.95

EASY ORIGAMI, John Montroll. Charming collection of 32 projects (hat, cup, pelican, piano, swan, many more) specially designed for the novice origami hobbyist. Clearly illustrated easy-to-follow instructions insure that even beginning papercrafters will achieve successful results. 48pp. 8¼ × 11. 27298-2 Pa. $2.95

THE COMPLETE BOOK OF BIRDHOUSE CONSTRUCTION FOR WOOD-WORKERS, Scott D. Campbell. Detailed instructions, illustrations, tables. Also data on bird habitat and instinct patterns. Bibliography. 3 tables. 63 illustrations in 15 figures. 48pp. 5¼ × 8½. 24407-5 Pa. $1.95

BLOOMINGDALE'S ILLUSTRATED 1886 CATALOG: Fashions, Dry Goods and Housewares, Bloomingdale Brothers. Famed merchants' extremely rare catalog depicting about 1,700 products: clothing, housewares, firearms, dry goods, jewelry, more. Invaluable for dating, identifying vintage items. Also, copyright-free graphics for artists, designers. Co-published with Henry Ford Museum & Green-field Village. 160pp. 8¼ × 11. 25780-0 Pa. $9.95

HISTORIC COSTUME IN PICTURES, Braun & Schneider. Over 1,450 costumed figures in clearly detailed engravings—from dawn of civilization to end of 19th century. Captions. Many folk costumes. 256pp. 8⅜ × 11¾. 23150-X Pa. $11.95

STICKLEY CRAFTSMAN FURNITURE CATALOGS, Gustav Stickley and L. & J. G. Stickley. Beautiful, functional furniture in two authentic catalogs from 1910. 594 illustrations, including 277 photos, show settles, rockers, armchairs, reclining chairs, bookcases, desks, tables. 183pp. 6½ × 9¼. 23838-5 Pa. $9.95

AMERICAN LOCOMOTIVES IN HISTORIC PHOTOGRAPHS: 1858 to 1949, Ron Ziel (ed.). A rare collection of 126 meticulously detailed official photographs, called "builder portraits," of American locomotives that majestically chronicle the rise of steam locomotive power in America. Introduction. Detailed captions. xi + 129pp. 9 × 12. 27393-8 Pa. $12.95

AMERICA'S LIGHTHOUSES: An Illustrated History, Francis Ross Holland, Jr. Delightfully written, profusely illustrated fact-filled survey of over 200 American lighthouses since 1716. History, anecdotes, technological advances, more. 240pp. 8 × 10¾. 25576-X Pa. $11.95

TOWARDS A NEW ARCHITECTURE, Le Corbusier. Pioneering manifesto by founder of "International School." Technical and aesthetic theories, views of industry, economics, relation of form to function, "mass-production split" and much more. Profusely illustrated. 320pp. 6⅛ × 9¼. (USO) 25023-7 Pa. $9.95

HOW THE OTHER HALF LIVES, Jacob Riis. Famous journalistic record, exposing poverty and degradation of New York slums around 1900, by major social reformer. 100 striking and influential photographs. 233pp. 10 × 7⅞.
22012-5 Pa $10.95

FRUIT KEY AND TWIG KEY TO TREES AND SHRUBS, William M. Harlow. One of the handiest and most widely used identification aids. Fruit key covers 120 deciduous and evergreen species; twig key 160 deciduous species. Easily used. Over 300 photographs. 126pp. 5⅜ × 8½. 20511-8 Pa. $3.95

COMMON BIRD SONGS, Dr. Donald J. Borror. Songs of 60 most common U.S. birds: robins, sparrows, cardinals, bluejays, finches, more—arranged in order of increasing complexity. Up to 9 variations of songs of each species.
Cassette and manual 99911-4 $8.95

ORCHIDS AS HOUSE PLANTS, Rebecca Tyson Northen. Grow cattleyas and many other kinds of orchids—in a window, in a case, or under artificial light. 63 illustrations. 148pp. 5⅜ × 8½. 23261-1 Pa. $4.95

MONSTER MAZES, Dave Phillips. Masterful mazes at four levels of difficulty. Avoid deadly perils and evil creatures to find magical treasures. Solutions for all 32 exciting illustrated puzzles. 48pp. 8¼ × 11. 26005-4 Pa. $2.95

MOZART'S DON GIOVANNI (DOVER OPERA LIBRETTO SERIES), Wolfgang Amadeus Mozart. Introduced and translated by Ellen H. Bleiler. Standard Italian libretto, with complete English translation. Convenient and thoroughly portable—an ideal companion for reading along with a recording or the performance itself. Introduction. List of characters. Plot summary. 121pp. 5¼ × 8½.
24944-1 Pa. $2.95

TECHNICAL MANUAL AND DICTIONARY OF CLASSICAL BALLET, Gail Grant. Defines, explains, comments on steps, movements, poses and concepts. 15-page pictorial section. Basic book for student, viewer. 127pp. 5⅜ × 8½.
21843-0 Pa. $4.95

BRASS INSTRUMENTS: Their History and Development, Anthony Baines. Authoritative, updated survey of the evolution of trumpets, trombones, bugles, cornets, French horns, tubas and other brass wind instruments. Over 140 illustrations and 48 music examples. Corrected and updated by author. New preface. Bibliography. 320pp. 5⅜ × 8½. 27574-4 Pa. $9.95

HOLLYWOOD GLAMOR PORTRAITS, John Kobal (ed.). 145 photos from 1926–49. Harlow, Gable, Bogart, Bacall; 94 stars in all. Full background on photographers, technical aspects. 160pp. 8⅜ × 11¼. 23352-9 Pa. $11.95

MAX AND MORITZ, Wilhelm Busch. Great humor classic in both German and English. Also 10 other works: "Cat and Mouse," "Plisch and Plumm," etc. 216pp. 5⅜ × 8½. 20181-3 Pa. $5.95

THE RAVEN AND OTHER FAVORITE POEMS, Edgar Allan Poe. Over 40 of the author's most memorable poems: "The Bells," "Ulalume," "Israfel," "To Helen," "The Conqueror Worm," "Eldorado," "Annabel Lee," many more. Alphabetic lists of titles and first lines. 64pp. 5³⁄₁₆ × 8¼. 26685-0 Pa. $1.00

SEVEN SCIENCE FICTION NOVELS, H. G. Wells. The standard collection of the great novels. Complete, unabridged. First Men in the Moon, Island of Dr. Moreau, War of the Worlds, Food of the Gods, Invisible Man, Time Machine, In the Days of the Comet. Total of 1,015pp. 5⅜ × 8½. (USO) 20264-X Clothbd. $29.95

AMULETS AND SUPERSTITIONS, E. A. Wallis Budge. Comprehensive discourse on origin, powers of amulets in many ancient cultures: Arab, Persian, Babylonian, Assyrian, Egyptian, Gnostic, Hebrew, Phoenician, Syriac, etc. Covers cross, swastika, crucifix, seals, rings, stones, etc. 584pp. 5⅜ × 8½. 23573-4 Pa. $12.95

RUSSIAN STORIES/PYCCKNE PACCKA3bl: A Dual-Language Book, edited by Gleb Struve. Twelve tales by such masters as Chekhov, Tolstoy, Dostoevsky, Pushkin, others. Excellent word-for-word English translations on facing pages, plus teaching and study aids, Russian/English vocabulary, biographical/critical introductions, more. 416pp. 5⅜ × 8½. 26244-8 Pa. $8.95

PHILADELPHIA THEN AND NOW: 60 Sites Photographed in the Past and Present, Kenneth Finkel and Susan Oyama. Rare photographs of City Hall, Logan Square, Independence Hall, Betsy Ross House, other landmarks juxtaposed with contemporary views. Captures changing face of historic city. Introduction. Captions. 128pp. 8¼ × 11. 25790-8 Pa. $9.95

AIA ARCHITECTURAL GUIDE TO NASSAU AND SUFFOLK COUNTIES, LONG ISLAND, The American Institute of Architects, Long Island Chapter, and the Society for the Preservation of Long Island Antiquities. Comprehensive, well-researched and generously illustrated volume brings to life over three centuries of Long Island's great architectural heritage. More than 240 photographs with authoritative, extensively detailed captions. 176pp. 8¼ × 11. 26946-9 Pa. $14.95

NORTH AMERICAN INDIAN LIFE: Customs and Traditions of 23 Tribes, Elsie Clews Parsons (ed.). 27 fictionalized essays by noted anthropologists examine religion, customs, government, additional facets of life among the Winnebago, Crow, Zuni, Eskimo, other tribes. 480pp. 6⅛ × 9¼. 27377-6 Pa. $10.95

FRANK LLOYD WRIGHT'S HOLLYHOCK HOUSE, Donald Hoffmann. Lavishly illustrated, carefully documented study of one of Wright's most controversial residential designs. Over 120 photographs, floor plans, elevations, etc. Detailed perceptive text by noted Wright scholar. Index. 128pp. 9¼ × 10¾.
27133-1 Pa. $11.95

THE MALE AND FEMALE FIGURE IN MOTION: 60 Classic Photographic Sequences, Eadweard Muybridge. 60 true-action photographs of men and women walking, running, climbing, bending, turning, etc., reproduced from rare 19th-century masterpiece. vi + 121pp. 9 × 12.
24745-7 Pa. $10.95

1001 QUESTIONS ANSWERED ABOUT THE SEASHORE, N. J. Berrill and Jacquelyn Berrill. Queries answered about dolphins, sea snails, sponges, starfish, fishes, shore birds, many others. Covers appearance, breeding, growth, feeding, much more. 305pp. 5¼ × 8¼.
23366-9 Pa. $7.95

GUIDE TO OWL WATCHING IN NORTH AMERICA, Donald S. Heintzelman. Superb guide offers complete data and descriptions of 19 species: barn owl, screech owl, snowy owl, many more. Expert coverage of owl-watching equipment, conservation, migrations and invasions, etc. Guide to observing sites. 84 illustrations. xiii + 193pp. 5⅜ × 8½.
27344-X Pa. $8.95

MEDICINAL AND OTHER USES OF NORTH AMERICAN PLANTS: A Historical Survey with Special Reference to the Eastern Indian Tribes, Charlotte Erichsen-Brown. Chronological historical citations document 500 years of usage of plants, trees, shrubs native to eastern Canada, northeastern U.S. Also complete identifying information. 343 illustrations. 544pp. 6½ × 9¼.
25951-X Pa. $12.95

STORYBOOK MAZES, Dave Phillips. 23 stories and mazes on two-page spreads: Wizard of Oz, Treasure Island, Robin Hood, etc. Solutions. 64pp. 8¼ × 11.
23628-5 Pa. $2.95

NEGRO FOLK MUSIC, U.S.A., Harold Courlander. Noted folklorist's scholarly yet readable analysis of rich and varied musical tradition. Includes authentic versions of over 40 folk songs. Valuable bibliography and discography. xi + 324pp. 5⅜ × 8½.
27350-4 Pa. $7.95

MOVIE-STAR PORTRAITS OF THE FORTIES, John Kobal (ed.). 163 glamor, studio photos of 106 stars of the 1940s: Rita Hayworth, Ava Gardner, Marlon Brando, Clark Gable, many more. 176pp. 8⅜ × 11¼.
23546-7 Pa. $11.95

BENCHLEY LOST AND FOUND, Robert Benchley. Finest humor from early 30s, about pet peeves, child psychologists, post office and others. Mostly unavailable elsewhere. 73 illustrations by Peter Arno and others. 183pp. 5⅜ × 8½.
22410-4 Pa. $5.95

YEKL and THE IMPORTED BRIDEGROOM AND OTHER STORIES OF YIDDISH NEW YORK, Abraham Cahan. Film Hester Street based on Yekl (1896). Novel, other stories among first about Jewish immigrants on N.Y.'s East Side. 240pp. 5⅜ × 8½.
22427-9 Pa. $6.95

SELECTED POEMS, Walt Whitman. Generous sampling from *Leaves of Grass*. Twenty-four poems include "I Hear America Singing," "Song of the Open Road," "I Sing the Body Electric," "When Lilacs Last in the Dooryard Bloom'd," "O Captain! My Captain!"—all reprinted from an authoritative edition. Lists of titles and first lines. 128pp. 5³⁄₁₆ × 8¼.
26878-0 Pa. $1.00

THE BEST TALES OF HOFFMANN, E. T. A. Hoffmann. 10 of Hoffmann's most important stories: "Nutcracker and the King of Mice," "The Golden Flowerpot," etc. 458pp. 5⅜ × 8½. 21793-0 Pa. $8.95

FROM FETISH TO GOD IN ANCIENT EGYPT, E. A. Wallis Budge. Rich detailed survey of Egyptian conception of "God" and gods, magic, cult of animals, Osiris, more. Also, superb English translations of hymns and legends. 240 illustrations. 545pp. 5⅜ × 8½. 25803-3 Pa. $11.95

FRENCH STORIES/CONTES FRANÇAIS: A Dual-Language Book, Wallace Fowlie. Ten stories by French masters, Voltaire to Camus: "Micromegas" by Voltaire; "The Atheist's Mass" by Balzac; "Minuet" by de Maupassant; "The Guest" by Camus, six more. Excellent English translations on facing pages. Also French-English vocabulary list, exercises, more. 352pp. 5⅜ × 8½. 26443-2 Pa. $8.95

CHICAGO AT THE TURN OF THE CENTURY IN PHOTOGRAPHS: 122 Historic Views from the Collections of the Chicago Historical Society, Larry A. Viskochil. Rare large-format prints offer detailed views of City Hall, State Street, the Loop, Hull House, Union Station, many other landmarks, circa 1904–1913. Introduction. Captions. Maps. 144pp. 9⅜ × 12¼. 24656-6 Pa. $12.95

OLD BROOKLYN IN EARLY PHOTOGRAPHS, 1865–1929, William Lee Younger. Luna Park, Gravesend race track, construction of Grand Army Plaza, moving of Hotel Brighton, etc. 157 previously unpublished photographs. 165pp. 8⅞ × 11¾. 23587-4 Pa. $13.95

THE MYTHS OF THE NORTH AMERICAN INDIANS, Lewis Spence. Rich anthology of the myths and legends of the Algonquins, Iroquois, Pawnees and Sioux, prefaced by an extensive historical and ethnological commentary. 36 illustrations. 480pp. 5⅜ × 8½. 25967-6 Pa. $8.95

AN ENCYCLOPEDIA OF BATTLES: Accounts of Over 1,560 Battles from 1479 B.C. to the Present, David Eggenberger. Essential details of every major battle in recorded history from the first battle of Megiddo in 1479 B.C. to Grenada in 1984. List of Battle Maps. New Appendix covering the years 1967–1984. Index. 99 illustrations. 544pp. 6½ × 9¼. 24913-1 Pa. $14.95

SAILING ALONE AROUND THE WORLD, Captain Joshua Slocum. First man to sail around the world, alone, in small boat. One of great feats of seamanship told in delightful manner. 67 illustrations. 294pp. 5⅜ × 8½. 20326-3 Pa. $5.95

ANARCHISM AND OTHER ESSAYS, Emma Goldman. Powerful, penetrating, prophetic essays on direct action, role of minorities, prison reform, puritan hypocrisy, violence, etc. 271pp. 5⅜ × 8½. 22484-8 Pa. $5.95

MYTHS OF THE HINDUS AND BUDDHISTS, Ananda K. Coomaraswamy and Sister Nivedita. Great stories of the epics; deeds of Krishna, Shiva, taken from puranas, Vedas, folk tales; etc. 32 illustrations. 400pp. 5⅜ × 8½. 21759-0 Pa. $9.95

BEYOND PSYCHOLOGY, Otto Rank. Fear of death, desire of immortality, nature of sexuality, social organization, creativity, according to Rankian system. 291pp. 5⅜ × 8½. 20485-5 Pa. $8.95

A THEOLOGICO-POLITICAL TREATISE, Benedict Spinoza. Also contains unfinished Political Treatise. Great classic on religious liberty, theory of government on common consent. R. Elwes translation. Total of 421pp. 5⅜ × 8½. 20249-6 Pa. $8.95

MY BONDAGE AND MY FREEDOM, Frederick Douglass. Born a slave, Douglass became outspoken force in antislavery movement. The best of Douglass' autobiographies. Graphic description of slave life. 464pp. 5⅜ × 8½.   22457-0 Pa. $8.95

FOLLOWING THE EQUATOR: A Journey Around the World, Mark Twain. Fascinating humorous account of 1897 voyage to Hawaii, Australia, India, New Zealand, etc. Ironic, bemused reports on peoples, customs, climate, flora and fauna, politics, much more. 197 illustrations. 720pp. 5⅜ × 8½.   26113-1 Pa. $15.95

THE PEOPLE CALLED SHAKERS, Edward D. Andrews. Definitive study of Shakers: origins, beliefs, practices, dances, social organization, furniture and crafts, etc. 33 illustrations. 351pp. 5⅜ × 8½.   21081-2 Pa. $8.95

THE MYTHS OF GREECE AND ROME, H. A. Guerber. A classic of mythology, generously illustrated, long prized for its simple, graphic, accurate retelling of the principal myths of Greece and Rome, and for its commentary on their origins and significance. With 64 illustrations by Michelangelo, Raphael, Titian, Rubens, Canova, Bernini and others. 480pp. 5⅜ × 8½.   27584-1 Pa. $9.95

PSYCHOLOGY OF MUSIC, Carl E. Seashore. Classic work discusses music as a medium from psychological viewpoint. Clear treatment of physical acoustics, auditory apparatus, sound perception, development of musical skills, nature of musical feeling, host of other topics. 88 figures. 408pp. 5⅜ × 8½. 21851-1 Pa. $9.95

THE PHILOSOPHY OF HISTORY, Georg W. Hegel. Great classic of Western thought develops concept that history is not chance but rational process, the evolution of freedom. 457pp. 5⅜ × 8½.   20112-0 Pa. $9.95

THE BOOK OF TEA, Kakuzo Okakura. Minor classic of the Orient: entertaining, charming explanation, interpretation of traditional Japanese culture in terms of tea ceremony. 94pp. 5⅜ × 8½.   20070-1 Pa. $3.95

LIFE IN ANCIENT EGYPT, Adolf Erman. Fullest, most thorough, detailed older account with much not in more recent books, domestic life, religion, magic, medicine, commerce, much more. Many illustrations reproduce tomb paintings, carvings, hieroglyphs, etc. 597pp. 5⅜ × 8½.   22632-8 Pa. $10.95

SUNDIALS, Their Theory and Construction, Albert Waugh. Far and away the best, most thorough coverage of ideas, mathematics concerned, types, construction, adjusting anywhere. Simple, nontechnical treatment allows even children to build several of these dials. Over 100 illustrations. 230pp. 5⅜ × 8½.   22947-5 Pa. $7.95

DYNAMICS OF FLUIDS IN POROUS MEDIA, Jacob Bear. For advanced students of ground water hydrology, soil mechanics and physics, drainage and irrigation engineering, and more. 335 illustrations. Exercises, with answers. 784pp. 6⅛ × 9¼.   65675-6 Pa. $19.95

SONGS OF EXPERIENCE: Facsimile Reproduction with 26 Plates in Full Color, William Blake. 26 full-color plates from a rare 1826 edition. Includes "The Tyger," "London," "Holy Thursday," and other poems. Printed text of poems. 48pp. 5¼ × 7. 24636-1 Pa. $4.95

OLD-TIME VIGNETTES IN FULL COLOR, Carol Belanger Grafton (ed.). Over 390 charming, often sentimental illustrations, selected from archives of Victorian graphics—pretty women posing, children playing, food, flowers, kittens and puppies, smiling cherubs, birds and butterflies, much more. All copyright-free. 48pp. 9¼ × 12¼.   27269-9 Pa. $5.95

PERSPECTIVE FOR ARTISTS, Rex Vicat Cole. Depth, perspective of sky and sea, shadows, much more, not usually covered. 391 diagrams, 81 reproductions of drawings and paintings. 279pp. 5⅜ × 8½. 22487-2 Pa. $6.95

DRAWING THE LIVING FIGURE, Joseph Sheppard. Innovative approach to artistic anatomy focuses on specifics of surface anatomy, rather than muscles and bones. Over 170 drawings of live models in front, back and side views, and in widely varying poses. Accompanying diagrams. 177 illustrations. Introduction. Index. 144pp. 8⅜ × 11¼. 26723-7 Pa. $8.95

GOTHIC AND OLD ENGLISH ALPHABETS: 100 Complete Fonts, Dan X. Solo. Add power, elegance to posters, signs, other graphics with 100 stunning copyright-free alphabets: Blackstone, Dolbey, Germania, 97 more—including many lower-case, numerals, punctuation marks. 104pp. 8⅛ × 11. 24695-7 Pa. $8.95

HOW TO DO BEADWORK, Mary White. Fundamental book on craft from simple projects to five-bead chains and woven works. 106 illustrations. 142pp. 5⅜ × 8. 20697-1 Pa. $4.95

THE BOOK OF WOOD CARVING, Charles Marshall Sayers. Finest book for beginners discusses fundamentals and offers 34 designs. "Absolutely first rate . . . well thought out and well executed."—E. J. Tangerman. 118pp. 7¾ × 10⅝. 23654-4 Pa. $5.95

ILLUSTRATED CATALOG OF CIVIL WAR MILITARY GOODS: Union Army Weapons, Insignia, Uniform Accessories, and Other Equipment, Schuyler, Hartley, and Graham. Rare, profusely illustrated 1846 catalog includes Union Army uniform and dress regulations, arms and ammunition, coats, insignia, flags, swords, rifles, etc. 226 illustrations. 160pp. 9 × 12. 24939-5 Pa. $10.95

WOMEN'S FASHIONS OF THE EARLY 1900s: An Unabridged Republication of "New York Fashions, 1909," National Cloak & Suit Co. Rare catalog of mail-order fashions documents women's and children's clothing styles shortly after the turn of the century. Captions offer full descriptions, prices. Invaluable resource for fashion, costume historians. Approximately 725 illustrations. 128pp. 8⅜ × 11¼. 27276-1 Pa. $11.95

THE 1912 AND 1915 GUSTAV STICKLEY FURNITURE CATALOGS, Gustav Stickley. With over 200 detailed illustrations and descriptions, these two catalogs are essential reading and reference materials and identification guides for Stickley furniture. Captions cite materials, dimensions and prices. 112pp. 6½ × 9¼. 26676-1 Pa. $9.95

EARLY AMERICAN LOCOMOTIVES, John H. White, Jr. Finest locomotive engravings from early 19th century: historical (1804–74), main-line (after 1870), special, foreign, etc. 147 plates. 142pp. 11⅜ × 8¼. 22772-3 Pa. $10.95

THE TALL SHIPS OF TODAY IN PHOTOGRAPHS, Frank O. Braynard. Lavishly illustrated tribute to nearly 100 majestic contemporary sailing vessels: Amerigo Vespucci, Clearwater, Constitution, Eagle, Mayflower, Sea Cloud, Victory, many more. Authoritative captions provide statistics, background on each ship. 190 black-and-white photographs and illustrations. Introduction. 128pp. 8⅞ × 11¾. 27163-3 Pa. $13.95

EARLY NINETEENTH-CENTURY CRAFTS AND TRADES, Peter Stockham (ed.). Extremely rare 1807 volume describes to youngsters the crafts and trades of the day: brickmaker, weaver, dressmaker, bookbinder, ropemaker, saddler, many more. Quaint prose, charming illustrations for each craft. 20 black-and-white line illustrations. 192pp. 4⅜ × 6. 27293-1 Pa. $4.95

VICTORIAN FASHIONS AND COSTUMES FROM HARPER'S BAZAR, 1867-1898, Stella Blum (ed.). Day costumes, evening wear, sports clothes, shoes, hats, other accessories in over 1,000 detailed engravings. 320pp. 9⅜ × 12¼. 22990-4 Pa. $13.95

GUSTAV STICKLEY, THE CRAFTSMAN, Mary Ann Smith. Superb study surveys broad scope of Stickley's achievement, especially in architecture. Design philosophy, rise and fall of the Craftsman empire, descriptions and floor plans for many Craftsman houses, more. 86 black-and-white halftones. 31 line illustrations. Introduction. 208pp. 6½ × 9¼. 27210-9 Pa. $9.95

THE LONG ISLAND RAIL ROAD IN EARLY PHOTOGRAPHS, Ron Ziel. Over 220 rare photos, informative text document origin (1844) and development of rail service on Long Island. Vintage views of early trains, locomotives, stations, passengers, crews, much more. Captions. 8⅞ × 11¾. 26301-0 Pa. $13.95

THE BOOK OF OLD SHIPS: From Egyptian Galleys to Clipper Ships, Henry B. Culver. Superb, authoritative history of sailing vessels, with 80 magnificent line illustrations. Galley, bark, caravel, longship, whaler, many more. Detailed, informative text on each vessel by noted naval historian. Introduction. 256pp. 5⅜ × 8½. 27332-6 Pa. $6.95

TEN BOOKS ON ARCHITECTURE, Vitruvius. The most important book ever written on architecture. Early Roman aesthetics, technology, classical orders, site selection, all other aspects. Morgan translation. 331pp. 5⅜ × 8½. 20645-9 Pa. $8.95

THE HUMAN FIGURE IN MOTION, Eadweard Muybridge. More than 4,500 stopped-action photos, in action series, showing undraped men, women, children jumping, lying down, throwing, sitting, wrestling, carrying, etc. 390pp. 7⅞ × 10⅝. 20204-6 Clothbd. $24.95

TREES OF THE EASTERN AND CENTRAL UNITED STATES AND CANADA, William M. Harlow. Best one-volume guide to 140 trees. Full descriptions, woodlore, range, etc. Over 600 illustrations. Handy size. 288pp. 4½ × 6⅜. 20395-6 Pa. $5.95

SONGS OF WESTERN BIRDS, Dr. Donald J. Borror. Complete song and call repertoire of 60 western species, including flycatchers, juncoes, cactus wrens, many more—includes fully illustrated booklet. Cassette and manual 99913-0 $8.95

GROWING AND USING HERBS AND SPICES, Milo Miloradovich. Versatile handbook provides all the information needed for cultivation and use of all the herbs and spices available in North America. 4 illustrations. Index. Glossary. 236pp. 5⅜ × 8½. 25058-X Pa. $6.95

BIG BOOK OF MAZES AND LABYRINTHS, Walter Shepherd. 50 mazes and labyrinths in all—classical, solid, ripple, and more—in one great volume. Perfect inexpensive puzzler for clever youngsters. Full solutions. 112pp. 8⅛ × 11. 22951-3 Pa. $4.95

PIANO TUNING, J. Cree Fischer. Clearest, best book for beginner, amateur. Simple repairs, raising dropped notes, tuning by easy method of flattened fifths. No previous skills needed. 4 illustrations. 201pp. 5⅜ × 8½. 23267-0 Pa. $5.95

A SOURCE BOOK IN THEATRICAL HISTORY, A. M. Nagler. Contemporary observers on acting, directing, make-up, costuming, stage props, machinery, scene design, from Ancient Greece to Chekhov. 611pp. 5⅜ × 8½. 20515-0 Pa. $11.95

THE COMPLETE NONSENSE OF EDWARD LEAR, Edward Lear. All nonsense limericks, zany alphabets, Owl and Pussycat, songs, nonsense botany, etc., illustrated by Lear. Total of 320pp. 5⅜ × 8½. (USO) 20167-8 Pa. $6.95

VICTORIAN PARLOUR POETRY: An Annotated Anthology, Michael R. Turner. 117 gems by Longfellow, Tennyson, Browning, many lesser-known poets. "The Village Blacksmith," "Curfew Must Not Ring Tonight," "Only a Baby Small," dozens more, often difficult to find elsewhere. Index of poets, titles, first lines. xxiii + 325pp. 5⅜ × 8¼. 27044-0 Pa. $8.95

DUBLINERS, James Joyce. Fifteen stories offer vivid, tightly focused observations of the lives of Dublin's poorer classes. At least one, "The Dead," is considered a masterpiece. Reprinted complete and unabridged from standard edition. 160pp. 5³⁄₁₆ × 8¼. 26870-5 Pa. $1.00

THE HAUNTED MONASTERY and THE CHINESE MAZE MURDERS, Robert van Gulik. Two full novels by van Gulik, set in 7th-century China, continue adventures of Judge Dee and his companions. An evil Taoist monastery, seemingly supernatural events; overgrown topiary maze hides strange crimes. 27 illustrations. 328pp. 5⅜ × 8½. 23502-5 Pa. $7.95

THE BOOK OF THE SACRED MAGIC OF ABRAMELIN THE MAGE, translated by S. MacGregor Mathers. Medieval manuscript of ceremonial magic. Basic document in Aleister Crowley, Golden Dawn groups. 268pp. 5⅜ × 8½. 23211-5 Pa. $8.95

NEW RUSSIAN-ENGLISH AND ENGLISH-RUSSIAN DICTIONARY, M. A. O'Brien. This is a remarkably handy Russian dictionary, containing a surprising amount of information, including over 70,000 entries. 366pp. 4½ × 6⅛. 20208-9 Pa. $9.95

HISTORIC HOMES OF THE AMERICAN PRESIDENTS, Second, Revised Edition, Irvin Haas. A traveler's guide to American Presidential homes, most open to the public, depicting and describing homes occupied by every American President from George Washington to George Bush. With visiting hours, admission charges, travel routes. 175 photographs. Index. 160pp. 8¼ × 11. 26751-2 Pa. $10.95

NEW YORK IN THE FORTIES, Andreas Feininger. 162 brilliant photographs by the well-known photographer, formerly with *Life* magazine. Commuters, shoppers, Times Square at night, much else from city at its peak. Captions by John von Hartz. 181pp. 9¼ × 10¾. 23585-8 Pa. $12.95

INDIAN SIGN LANGUAGE, William Tomkins. Over 525 signs developed by Sioux and other tribes. Written instructions and diagrams. Also 290 pictographs. 111pp. 6⅛ × 9¼. 22029-X Pa. $3.50

ANATOMY: A Complete Guide for Artists, Joseph Sheppard. A master of figure drawing shows artists how to render human anatomy convincingly. Over 460 illustrations. 224pp. 8⅜ × 11¼. 27279-6 Pa. $10.95

MEDIEVAL CALLIGRAPHY: Its History and Technique, Marc Drogin. Spirited history, comprehensive instruction manual covers 13 styles (ca. 4th century thru 15th). Excellent photographs; directions for duplicating medieval techniques with modern tools. 224pp. 8⅜ × 11¼. 26142-5 Pa. $11.95

DRIED FLOWERS: How to Prepare Them, Sarah Whitlock and Martha Rankin. Complete instructions on how to use silica gel, meal and borax, perlite aggregate, sand and borax, glycerine and water to create attractive permanent flower arrangements. 12 illustrations. 32pp. 5⅜ × 8½. 21802-3 Pa. $1.00

EASY-TO-MAKE BIRD FEEDERS FOR WOODWORKERS, Scott D. Campbell. Detailed, simple-to-use guide for designing, constructing, caring for and using feeders. Text, illustrations for 12 classic and contemporary designs. 96pp. 5⅜ × 8½. 25847-5 Pa. $2.95

OLD-TIME CRAFTS AND TRADES, Peter Stockham. An 1807 book created to teach children about crafts and trades open to them as future careers. It describes in detailed, nontechnical terms 24 different occupations, among them coachmaker, gardener, hairdresser, lacemaker, shoemaker, wheelwright, copper-plate printer, milliner, trunkmaker, merchant and brewer. Finely detailed engravings illustrate each occupation. 192pp. 4⅝ × 6. 27398-9 Pa. $4.95

THE HISTORY OF UNDERCLOTHES, C. Willett Cunnington and Phyllis Cunnington. Fascinating, well-documented survey covering six centuries of English undergarments, enhanced with over 100 illustrations: 12th-century laced-up bodice, footed long drawers (1795), 19th-century bustles, 19th-century corsets for men, Victorian "bust improvers," much more. 272pp. 5⅝ × 8¼. 27124-2 Pa. $9.95

ARTS AND CRAFTS FURNITURE: The Complete Brooks Catalog of 1912, Brooks Manufacturing Co. Photos and detailed descriptions of more than 150 now very collectible furniture designs from the Arts and Crafts movement depict davenports, settees, buffets, desks, tables, chairs, bedsteads, dressers and more, all built of solid, quarter-sawed oak. Invaluable for students and enthusiasts of antiques, Americana and the decorative arts. 80pp. 6½ × 9¼. 27471-3 Pa. $7.95

HOW WE INVENTED THE AIRPLANE: An Illustrated History, Orville Wright. Fascinating firsthand account covers early experiments, construction of planes and motors, first flights, much more. Introduction and commentary by Fred C. Kelly. 76 photographs. 96pp. 8¼ × 11. 25662-6 Pa. $8.95

THE ARTS OF THE SAILOR: Knotting, Splicing and Ropework, Hervey Garrett Smith. Indispensable shipboard reference covers tools, basic knots and useful hitches; handsewing and canvas work, more. Over 100 illustrations. Delightful reading for sea lovers. 256pp. 5⅜ × 8½. 26440-8 Pa. $7.95

FRANK LLOYD WRIGHT'S FALLINGWATER: The House and Its History, Second, Revised Edition, Donald Hoffmann. A total revision—both in text and illustrations—of the standard document on Fallingwater, the boldest, most personal architectural statement of Wright's mature years, updated with valuable new material from the recently opened Frank Lloyd Wright Archives. "Fascinating"—*The New York Times*. 116 illustrations. 128pp. 9¼ × 10¾. 27430-6 Pa. $10.95

PHOTOGRAPHIC SKETCHBOOK OF THE CIVIL WAR, Alexander Gardner. 100 photos taken on field during the Civil War. Famous shots of Manassas, Harper's Ferry, Lincoln, Richmond, slave pens, etc. 244pp. 10⅝ × 8¼.
22731-6 Pa. $9.95

FIVE ACRES AND INDEPENDENCE, Maurice G. Kains. Great back-to-the-land classic explains basics of self-sufficient farming. The one book to get. 95 illustrations. 397pp. 5⅜ × 8½.
20974-1 Pa. $7.95

SONGS OF EASTERN BIRDS, Dr. Donald J. Borror. Songs and calls of 60 species most common to eastern U.S.: warblers, woodpeckers, flycatchers, thrushes, larks, many more in high-quality recording.
Cassette and manual 99912-2 $8.95

A MODERN HERBAL, Margaret Grieve. Much the fullest, most exact, most useful compilation of herbal material. Gigantic alphabetical encyclopedia, from aconite to zedoary, gives botanical information, medical properties, folklore, economic uses, much else. Indispensable to serious reader. 161 illustrations. 888pp. 6½ × 9¼. 2-vol. set. (USO)
Vol. I: 22798-7 Pa. $9.95
Vol. II: 22799-5 Pa. $9.95

HIDDEN TREASURE MAZE BOOK, Dave Phillips. Solve 34 challenging mazes accompanied by heroic tales of adventure. Evil dragons, people-eating plants, bloodthirsty giants, many more dangerous adversaries lurk at every twist and turn. 34 mazes, stories, solutions. 48pp. 8¼ × 11.
24566-7 Pa. $2.95

LETTERS OF W. A. MOZART, Wolfgang A. Mozart. Remarkable letters show bawdy wit, humor, imagination, musical insights, contemporary musical world; includes some letters from Leopold Mozart. 276pp. 5⅜ × 8½.
22859-2 Pa. $7.95

BASIC PRINCIPLES OF CLASSICAL BALLET, Agrippina Vaganova. Great Russian theoretician, teacher explains methods for teaching classical ballet. 118 illustrations. 175pp. 5⅜ × 8½.
22036-2 Pa. $4.95

THE JUMPING FROG, Mark Twain. Revenge edition. The original story of The Celebrated Jumping Frog of Calaveras County, a hapless French translation, and Twain's hilarious "retranslation" from the French. 12 illustrations. 66pp. 5⅜ × 8½.
22686-7 Pa. $3.95

BEST REMEMBERED POEMS, Martin Gardner (ed.). The 126 poems in this superb collection of 19th- and 20th-century British and American verse range from Shelley's "To a Skylark" to the impassioned "Renascence" of Edna St. Vincent Millay and to Edward Lear's whimsical "The Owl and the Pussycat." 224pp. 5⅜ × 8½.
27165-X Pa. $4.95

COMPLETE SONNETS, William Shakespeare. Over 150 exquisite poems deal with love, friendship, the tyranny of time, beauty's evanescence, death and other themes in language of remarkable power, precision and beauty. Glossary of archaic terms. 80pp. 5³⁄₁₆ × 8¼.
26686-9 Pa. $1.00

BODIES IN A BOOKSHOP, R. T. Campbell. Challenging mystery of blackmail and murder with ingenious plot and superbly drawn characters. In the best tradition of British suspense fiction. 192pp. 5⅜ × 8½.
24720-1 Pa. $5.95

THE WIT AND HUMOR OF OSCAR WILDE, Alvin Redman (ed.). More than 1,000 ripostes, paradoxes, wisecracks: Work is the curse of the drinking classes; I can resist everything except temptation; etc. 258pp. 5⅜ × 8½. 20602-5 Pa. $5.95

SHAKESPEARE LEXICON AND QUOTATION DICTIONARY, Alexander Schmidt. Full definitions, locations, shades of meaning in every word in plays and poems. More than 50,000 exact quotations. 1,485pp. 6½ × 9¼. 2-vol. set.
Vol. I: 22726-X Pa. $16.95
Vol. 2: 22727-8 Pa. $15.95

SELECTED POEMS, Emily Dickinson. Over 100 best-known, best-loved poems by one of America's foremost poets, reprinted from authoritative early editions. No comparable edition at this price. Index of first lines. 64pp. 5³⁄₁₆ × 8¼.
26466-1 Pa. $1.00

CELEBRATED CASES OF JUDGE DEE (DEE GOONG AN), translated by Robert van Gulik. Authentic 18th-century Chinese detective novel; Dee and associates solve three interlocked cases. Led to van Gulik's own stories with same characters. Extensive introduction. 9 illustrations. 237pp. 5⅜ × 8½.
23337-5 Pa. $6.95

THE MALLEUS MALEFICARUM OF KRAMER AND SPRENGER, translated by Montague Summers. Full text of most important witchhunter's "bible," used by both Catholics and Protestants. 278pp. 6⅝ × 10. 22802-9 Pa. $11.95

SPANISH STORIES/CUENTOS ESPAÑOLES: A Dual-Language Book, Angel Flores (ed.). Unique format offers 13 great stories in Spanish by Cervantes, Borges, others. Faithful English translations on facing pages. 352pp. 5⅜ × 8½.
25399-6 Pa. $8.95

THE CHICAGO WORLD'S FAIR OF 1893: A Photographic Record, Stanley Appelbaum (ed.). 128 rare photos show 200 buildings, Beaux-Arts architecture, Midway, original Ferris Wheel, Edison's kinetoscope, more. Architectural emphasis; full text. 116pp. 8¼ × 11. 23990-X Pa. $9.95

OLD QUEENS, N.Y., IN EARLY PHOTOGRAPHS, Vincent F. Seyfried and William Asadorian. Over 160 rare photographs of Maspeth, Jamaica, Jackson Heights, and other areas. Vintage views of DeWitt Clinton mansion, 1939 World's Fair and more. Captions. 192pp. 8⅞ × 11. 26358-4 Pa. $12.95

CAPTURED BY THE INDIANS: 15 Firsthand Accounts, 1750–1870, Frederick Drimmer. Astounding true historical accounts of grisly torture, bloody conflicts, relentless pursuits, miraculous escapes and more, by people who lived to tell the tale. 384pp. 5⅜ × 8½. 24901-8 Pa. $8.95

THE WORLD'S GREAT SPEECHES, Lewis Copeland and Lawrence W. Lamm (eds.). Vast collection of 278 speeches of Greeks to 1970. Powerful and effective models; unique look at history. 842pp. 5⅜ × 8½. 20468-5 Pa. $14.95

THE BOOK OF THE SWORD, Sir Richard F. Burton. Great Victorian scholar/adventurer's eloquent, erudite history of the "queen of weapons"—from prehistory to early Roman Empire. Evolution and development of early swords, variations (sabre, broadsword, cutlass, scimitar, etc.), much more. 336pp. 6⅛ × 9¼. 25434-8 Pa. $8.95

AUTOBIOGRAPHY: The Story of My Experiments with Truth, Mohandas K. Gandhi. Boyhood, legal studies, purification, the growth of the Satyagraha (nonviolent protest) movement. Critical, inspiring work of the man responsible for the freedom of India. 480pp. 5⅜ × 8½. (USO)   24593-4 Pa. $8.95

CELTIC MYTHS AND LEGENDS, T. W. Rolleston. Masterful retelling of Irish and Welsh stories and tales. Cuchulain, King Arthur, Deirdre, the Grail, many more. First paperback edition. 58 full-page illustrations. 512pp. 5⅜ × 8½.
26507-2 Pa. $9.95

THE PRINCIPLES OF PSYCHOLOGY, William James. Famous long course complete, unabridged. Stream of thought, time perception, memory, experimental methods; great work decades ahead of its time. 94 figures. 1,391pp. 5⅜ × 8½. 2-vol. set.
Vol. I: 20381-6 Pa. $12.95
Vol. II: 20382-4 Pa. $12.95

THE WORLD AS WILL AND REPRESENTATION, Arthur Schopenhauer. Definitive English translation of Schopenhauer's life work, correcting more than 1,000 errors, omissions in earlier translations. Translated by E. F. J. Payne. Total of 1,269pp. 5⅜ × 8½. 2-vol. set.   Vol. 1: 21761-2 Pa. $11.95
Vol. 2: 21762-0 Pa. $11.95

MAGIC AND MYSTERY IN TIBET, Madame Alexandra David-Neel. Experiences among lamas, magicians, sages, sorcerers, Bonpa wizards. A true psychic discovery. 32 illustrations. 321pp. 5⅜ × 8½. (USO)   22682-4 Pa. $8.95

THE EGYPTIAN BOOK OF THE DEAD, E. A. Wallis Budge. Complete reproduction of Ani's papyrus, finest ever found. Full hieroglyphic text, interlinear transliteration, word-for-word translation, smooth translation. 533pp. 6½ × 9¼.
21866-X Pa. $9.95

MATHEMATICS FOR THE NONMATHEMATICIAN, Morris Kline. Detailed, college-level treatment of mathematics in cultural and historical context, with numerous exercises. Recommended Reading Lists. Tables. Numerous figures. 641pp. 5⅜ × 8½.   24823-2 Pa. $11.95

THEORY OF WING SECTIONS: Including a Summary of Airfoil Data, Ira H. Abbott and A. E. von Doenhoff. Concise compilation of subsonic aerodynamic characteristics of NACA wing sections, plus description of theory. 350pp. of tables. 693pp. 5⅜ × 8½.   60586-8 Pa. $14.95

THE RIME OF THE ANCIENT MARINER, Gustave Doré, S. T. Coleridge. Doré's finest work; 34 plates capture moods, subtleties of poem. Flawless full-size reproductions printed on facing pages with authoritative text of poem. "Beautiful. Simply beautiful."—Publisher's Weekly. 77pp. 9¼ × 12.   22305-1 Pa. $6.95

NORTH AMERICAN INDIAN DESIGNS FOR ARTISTS AND CRAFTS-PEOPLE, Eva Wilson. Over 360 authentic copyright-free designs adapted from Navajo blankets, Hopi pottery, Sioux buffalo hides, more. Geometrics, symbolic figures, plant and animal motifs, etc. 128pp. 8⅜ × 11. (EUK)   25341-4 Pa. $7.95

SCULPTURE: Principles and Practice, Louis Slobodkin. Step-by-step approach to clay, plaster, metals, stone; classical and modern. 253 drawings, photos. 255pp. 8⅛ × 11.   22960-2 Pa. $10.95

THE INFLUENCE OF SEA POWER UPON HISTORY, 1660–1783, A. T. Mahan. Influential classic of naval history and tactics still used as text in war colleges. First paperback edition. 4 maps. 24 battle plans. 640pp. 5⅜ × 8½.
25509-3 Pa. $12.95

THE STORY OF THE TITANIC AS TOLD BY ITS SURVIVORS, Jack Winocour (ed.). What it was really like. Panic, despair, shocking inefficiency, and a little heroism. More thrilling than any fictional account. 26 illustrations. 320pp. 5⅜ × 8½. 20610-6 Pa. **$8.95**

FAIRY AND FOLK TALES OF THE IRISH PEASANTRY, William Butler Yeats (ed.). Treasury of 64 tales from the twilight world of Celtic myth and legend: "The Soul Cages," "The Kildare Pooka," "King O'Toole and his Goose," many more. Introduction and Notes by W. B. Yeats. 352pp. 5⅜ × 8½. 26941-8 Pa. **$8.95**

BUDDHIST MAHAYANA TEXTS, E. B. Cowell and Others (eds.). Superb, accurate translations of basic documents in Mahayana Buddhism, highly important in history of religions. The Buddha-karita of Asvaghosha, Larger Sukhavativyuha, more. 448pp. 5⅜ × 8½. , 25552-2 Pa. $9.95

ONE TWO THREE . . . INFINITY: Facts and Speculations of Science, George Gamow. Great physicist's fascinating, readable overview of contemporary science: number theory, relativity, fourth dimension, entropy, genes, atomic structure, much more. 128 illustrations. Index. 352pp. 5⅜ × 8½. 25664-2 Pa. **$8.95**

ENGINEERING IN HISTORY, Richard Shelton Kirby, et al. Broad, nontechnical survey of history's major technological advances: birth of Greek science, industrial revolution, electricity and applied science, 20th-century automation, much more. 181 illustrations. ". . . excellent . . ."—Isis. Bibliography. vii + 530pp. 5⅜ × 8¼.
26412-2 Pa. $14.95